Learn Python From an Expert

© 2023 by Edson L P Camacho

Dedication

It was a battle to write a book of this size and with this expert content, to deliver the best there is for Python language users.

My family was my support, I want to thank my wife Vanessa, my son Giovanni and my God for giving me strength to complete this project.

I offer it to all my readers, and I thank everyone for trusting the information contained in this book, as support for the solutions they may eventually have.

This is a great journey, only the strong and those who don't give up make it to the end, I wish you all success!

One day the prophet Isaiah said...

"All men are like grass and all their glory is like the flowers of the field... The grass withers and the flowers fall, but the Word of our God stands forever."

Isaiah 40: 7-8

Edson L P Camacho

Table of Contents

○ Important introduction to Python

Why use the Python language

There are many reasons why Python is a popular and widely used language for various applications, such as:

Easy to learn: Python has a simple and intuitive syntax that makes it easy to learn for beginners. The language emphasizes readability and clarity, which makes it easier to understand and maintain code.

Versatility: Python can be used for a wide range of applications, including web development, scientific computing, artificial intelligence, machine learning, data analysis, and more.

Large community and extensive libraries: Python has a large and active community of developers, which means there is a wealth of resources and support available online. There are also many libraries and frameworks available that make it easier to develop applications in Python.

Cross-platform compatibility: Python can be run on various operating systems, including Windows, macOS, and Linux, making it a versatile language for development.

Rapid prototyping: Python's simplicity and ease of use make it a popular choice for rapid prototyping and development of minimum viable products (MVPs).

High demand and career opportunities: Python is one of the most in-demand programming languages in the job market, and it offers various career opportunities for developers, data scientists, and AI/ML professionals.

Overall, Python is a powerful and flexible language that is well-suited for a wide range of applications and industries. Its simplicity, versatility, and large community make it an excellent choice for both beginners and experienced developers.

Python is considered easy to learn for several reasons:

1. Simple and readable syntax: Python has a simple and intuitive syntax that emphasizes readability and reduces the complexity of the code. The language uses plain English words and a minimalist design that makes it easy to understand and write code.

2. Fewer lines of code: Python requires fewer lines of code to accomplish the same task as other programming languages, which reduces the time and effort required to write, test, and maintain code.

3. Extensive documentation and community support: Python has a large and active community of developers who contribute to its development and maintenance. This community provides extensive documentation, tutorials, and resources that make it easier for beginners to learn and use Python.

4. Interactive mode: Python allows developers to enter code interactively and see the results immediately, which provides instant feedback and makes it easier to experiment and learn.

5. Versatility: Python can be used for a wide range of applications, from web development and data analysis to machine learning and scientific computing. Its versatility makes it easier for developers to find applications and projects that interest them.

Overall, Python's simplicity, readability, and versatility make it a popular choice for beginners and experienced developers alike. The language's ease of use and extensive community support make it easier for developers to learn, use, and master Python.

Is Python versatility?

Yes, Python is a versatile programming language that can be used for a wide range of applications. Python's versatility comes from the fact that it is a general-purpose language that can be used for many different purposes, including:

Web development: Python is commonly used for web development, with popular frameworks such as Django and Flask.

Data analysis and visualization: Python has a wide range of libraries and tools for data analysis and visualization, such as NumPy, Pandas, and Matplotlib.

Scientific computing: Python is widely used in scientific computing, with libraries such as SciPy, SymPy, and Biopython.

Artificial intelligence and machine learning: Python is one of the most popular languages for machine learning and AI, with libraries such as TensorFlow, Keras, and PyTorch.

Desktop applications: Python can be used to develop desktop applications using frameworks such as PyQt and PyGTK.

Game development: Python can be used for game development, with libraries such as Pygame.

Automation and scripting: Python is commonly used for automation and scripting tasks, such as system administration, network programming, and web scraping.

Overall, Python's versatility makes it a popular choice for a wide range of applications and industries, from startups and small businesses to large enterprises and scientific research institutions.

Python with large community and extensive libraries

Python has a large and active community of developers and users, which is one of its biggest strengths. The community is constantly contributing to the development and improvement of Python, and there is a wealth of resources and support available online.

In addition to its community, Python also has an extensive library of modules and packages that can be used for various purposes. These libraries include:

1. NumPy: a library for numerical computing and scientific computing.
2. Pandas: a library for data manipulation and analysis.
3. Matplotlib: a library for creating visualizations and plots.
4. Scikit-learn: a library for machine learning and data mining.
5. TensorFlow: a library for deep learning and neural networks.
6. Flask and Django: web development frameworks that are widely used for building web applications.
7. Pygame: a library for game development.
8. BeautifulSoup: a library for web scraping.
9. Requests: a library for making HTTP requests.
10. Pytest: a testing framework for Python.

These libraries and packages make it easier for developers to build complex applications and solve challenging problems. They also help to reduce the amount of time and effort required to develop and maintain code.

Overall, Python's large community and extensive library of modules and packages make it a powerful and versatile language that is well-suited for a wide range of applications and industries.

12

Where is python used nowadays?

Python is used in a wide range of industries and applications. Some of the most common uses of Python include:

1. Web development: Python is widely used for web development, with popular frameworks such as Django, Flask, and Pyramid.
2. Data analysis and visualization: Python has a variety of libraries and tools for data analysis and visualization, including Pandas, NumPy, Matplotlib, and Seaborn.
3. Machine learning and artificial intelligence: Python is one of the most popular languages for machine learning and AI, with popular libraries such as TensorFlow, Keras, and Scikit-learn.
4. Scientific computing: Python is widely used in scientific computing, with libraries such as SciPy, SymPy, and Biopython.
5. Desktop applications: Python can be used to develop desktop applications using frameworks such as PyQt and wxPython.
6. Game development: Python can be used for game development, with libraries such as Pygame.
7. Automation and scripting: Python is commonly used for automation and scripting tasks, such as system administration, network programming, and web scraping.
8. Education: Python is often used as an introductory programming language in education, as its simplicity and readability make it easy for beginners to learn.
9. Finance: Python is widely used in finance for data analysis, algorithmic trading, and risk management.
10. DevOps: Python is used for automation and orchestration in DevOps, with popular tools such as Ansible, Salt, and Fabric.

Overall, Python's versatility and ease of use make it a popular choice for developers in many different industries and applications. Its large community and extensive library of modules and packages also make it a powerful language for solving complex problems and building robust applications.

What can we do with the Python language?

Python is a versatile programming language that can be used for a wide variety of tasks and applications. Here are some examples of what you can do with Python:

1. Web development: Python can be used to create web applications using popular frameworks such as Django, Flask, and Pyramid.
2. Data analysis and visualization: Python has a range of libraries and tools for data analysis and visualization, such as Pandas, NumPy, Matplotlib, and Seaborn.

3. Machine learning and artificial intelligence: Python is one of the most popular languages for machine learning and AI, with libraries such as TensorFlow, Keras, Scikit-learn, and PyTorch.Scientific computing: Python is widely used in scientific computing, with libraries such as SciPy, SymPy, and Biopython.
4. Desktop applications: Python can be used to develop desktop applications using frameworks such as PyQt and wxPython.
5. Game development: Python can be used for game development, with libraries such as Pygame.
6. Automation and scripting: Python is commonly used for automation and scripting tasks, such as system administration, network programming, and web scraping.
7. Education: Python is often used as an introductory programming language in education, as its simplicity and readability make it easy for beginners to learn.
8. Finance: Python is widely used in finance for data analysis, algorithmic trading, and risk management.
9. DevOps: Python is used for automation and orchestration in DevOps, with popular tools such as Ansible, Salt, and Fabric.

Overall, Python is a powerful language that can be used for a wide range of tasks and applications. Its versatility and ease of use make it a popular choice for developers in many different industries and domains.

What is the downside of using Python?

While Python is a powerful and versatile language, it does have some limitations and drawbacks. Here are some potential downsides to using Python:

Performance: Python is an interpreted language, which means it can be slower than compiled languages like C or Java. This can make it less suitable for certain types of high-performance applications, such as real-time systems or resource-intensive tasks.

GIL: Python has a Global Interpreter Lock (GIL), which can limit its ability to take full advantage of multi-core processors. This can make it less suitable for certain types of CPU-bound tasks.

Memory management: Python uses automatic memory management, which can lead to performance issues if memory usage is not carefully managed. This can be a problem for applications that need to process large amounts of data.

Version compatibility: Python has multiple versions, and not all libraries or frameworks are compatible with every version. This can create compatibility issues and make it difficult to maintain code over time.

Packaging: Python's packaging system can be complex and difficult to manage, especially for large projects with many dependencies.

Security: Python is a popular language, which can make it a target for security vulnerabilities and attacks.

Overall, while Python has many advantages, it may not be the best choice for every application or use case. Developers should carefully consider their requirements and constraints before deciding to use Python.

What does Python do?

Python in System Programming

Python is a versatile programming language used for various tasks, including system programming. Python's popularity stems from its flexibility and readability, making it an ideal language for system programming. In this article, we'll explore what Python does in system programming and its benefits.

Overview of System Programming

System programming involves creating software that interacts with the hardware and operating system of a computer. It's a low-level task that requires developers to have a good understanding of the computer's architecture and how it interacts with the operating system. System programming is often used to create system utilities, device drivers, and operating systems.

Python in System Programming

Python is a high-level language that has been gaining popularity in system programming. Despite being a high-level language, it still offers low-level programming capabilities that make it an ideal language for system programming. Python's extensive standard library provides developers with access to a wide range of system functionalities, making it easier to write system programs.

Python's Benefits in System Programming

Python has several benefits in system programming, including:

Easy to Learn: Python is an easy-to-learn language that can be used by both novice and experienced developers. It's a high-level language with simple syntax and semantics that make it easy to read and write code.

Cross-Platform Compatibility: Python is a cross-platform language that can run on different operating systems, including Windows, Linux, and macOS. This makes it easier to develop system programs that can run on different platforms without the need for additional software.

Large Community: Python has a large community of developers who contribute to its development and share their knowledge and expertise. This community provides support and resources to developers, making it easier to find help and solutions to problems.

Extensive Libraries: Python has an extensive standard library that provides developers with access to various functionalities for system programming. These libraries make it easier to write system programs without the need for additional code.

Conclusion

Python has become a popular language in system programming due to its flexibility, ease of use, and extensive libraries. Python's ability to run on different operating systems and its large community of developers make it an ideal language for system programming. Its simplicity and low-level capabilities make it an easy language to learn for both novice and experienced developers. As the demand for system programming continues to grow, Python's role in this field is expected to increase.

Python in Digital Games

Python is a popular programming language used for a wide range of applications. One of the areas where Python has made a significant impact is in digital games development. Python is used to create games of all types, from simple 2D platformers to complex 3D simulations.

The popularity of Python in the digital games industry can be attributed to its simplicity, versatility, and flexibility. Python's syntax is easy to read and understand, making it accessible to both new and experienced developers. Python's ability to integrate with other programming languages and software tools also makes it a popular choice for game development.

Python in Data Mining

Data mining is the process of analyzing and extracting meaningful patterns and insights from large datasets. Python is a powerful tool for data mining and has become a go-to language for many data scientists and analysts.

Python's simplicity, ease of use, and powerful libraries such as NumPy, Pandas, and Scikit-learn, make it ideal for data mining tasks. These libraries provide a range of data analysis and visualization tools that allow developers to process and analyze data quickly and efficiently.

Python in Robotics

Python is also widely used in robotics programming. Robotics is the branch of technology that deals with the design, construction, operation, and application of robots. Python's simplicity and versatility make it an ideal language for robotics programming.

Python's ease of use and readability make it a popular choice for beginners looking to learn robotics programming. Python's extensive library ecosystem also provides developers with a range of tools and frameworks for robotics programming, such as PyRobot, ROS, and TensorFlow.

Python in Image Processing

Image processing is the process of manipulating and analyzing digital images using algorithms. Python is a popular language for image processing due to its simplicity, powerful libraries, and versatility.

Python's libraries such as OpenCV, Pillow, and Scikit-image provide developers with a range of tools for image processing, such as image filtering, edge detection, and object recognition. These tools can be used in a variety of applications, from medical imaging to facial recognition.

Conclusion

Python is a versatile and powerful programming language that has made significant contributions to many areas of technology, including digital games, data mining, robotics, and image processing. Python's simplicity, ease of use, and powerful libraries have made it a popular choice for developers looking to create powerful and efficient applications in these areas. Whether you're a beginner or an experienced developer, Python has something to offer for everyone.

• Python and Database Programming

Python is a popular programming language that has gained a lot of popularity in recent years due to its simplicity, flexibility, and ease of use. One of the areas in which Python has made significant inroads is database programming. In this article, we will discuss what Python does with database programming and how it is used to develop robust, scalable, and efficient database applications.

Python and Database Connectivity

Python provides an extensive library of modules for database connectivity. These modules allow programmers to connect to a wide range of databases, such as MySQL, PostgreSQL, Oracle, and Microsoft SQL Server, among others. The modules provide an API for performing common database operations such as querying, updating, and inserting data. These modules provide a convenient way of interacting with databases from within Python programs, without the need for writing complex SQL statements.

Python and Data Modeling

Python is also widely used for data modeling in database programming. Python provides an object-relational mapping (ORM) framework that allows developers to map object-oriented data models to relational databases. This framework simplifies the development of database applications by eliminating the need for writing complex SQL statements. The ORM framework also provides features such as lazy loading, caching, and optimistic locking, which can help improve the performance of database applications.

Python and Database Administration

Python is also used for database administration tasks. Python provides modules for managing databases, such as creating and deleting tables, backing up and restoring databases, and managing user accounts. These modules provide a convenient way of automating database administration tasks, which can help reduce errors and increase productivity.

Python and Data Analysis

Python is widely used for data analysis in database programming. Python provides powerful libraries for data analysis, such as NumPy, Pandas, and Matplotlib, among others. These libraries provide a convenient way of analyzing large amounts of data and generating meaningful insights. Python can also be used for data mining and machine learning tasks, which can help organizations gain a competitive advantage by predicting trends and making informed decisions.

Python and Big Data

Python is also used for big data applications in database programming. Python provides libraries such as PySpark and Dask, which allow developers to work with big data technologies

such as Hadoop, Spark, and NoSQL databases. These libraries provide a convenient way of processing large amounts of data in parallel, which can help improve the performance of big data applications.

Conclusion

Python is a versatile programming language that is widely used in database programming. Python provides extensive libraries for database connectivity, data modeling, database administration, data analysis, and big data applications. Python's simplicity, flexibility, and ease of use make it an ideal choice for developing robust, scalable, and efficient database applications.

Python and GUIs

Python is a general-purpose programming language that is widely used for web development, data analysis, and scientific computing. It has also proven to be a popular choice for building graphical user interfaces (GUIs) due to its simplicity and ease of use. Python has several GUI toolkits available that make it easy to create GUI applications for various platforms and operating systems. In this article, we will explore how Python is used in GUI programming and some of the popular Python GUI toolkits.

GUI Toolkits in Python

There are several GUI toolkits available in Python, each with its own strengths and weaknesses. Here are some of the most popular ones:

1. Tkinter: Tkinter is the standard GUI toolkit that comes bundled with Python. It is a lightweight toolkit and is very easy to use. Tkinter provides a wide range of widgets, such as buttons, labels, and text boxes, which can be used to create simple to moderately complex GUI applications.
2. PyQt: PyQt is a set of Python bindings for the popular C++ Qt toolkit. It is a powerful and feature-rich toolkit that provides a wide range of widgets, such as buttons, menus, and text boxes, as well as support for advanced features like drag and drop, database integration, and multimedia.
3. wxPython: wxPython is a Python wrapper for the C++ wxWidgets toolkit. It is a cross-platform toolkit that provides native look and feel for each platform. wxPython provides a wide range of widgets and advanced features like database integration, drag and drop, and multimedia.
4. Kivy: Kivy is an open-source Python library for developing mobile and desktop applications. It provides a comprehensive set of widgets, such as buttons, text boxes, and sliders, as well as support for advanced features like multitouch, graphics, and animations.

GUI Programming with Tkinter

As mentioned earlier, Tkinter is the standard GUI toolkit that comes bundled with Python. It is a lightweight toolkit and is very easy to use. Here is a simple example of how to create a GUI application using Tkinter:

```
import tkinter as tk

def say_hello():
    print("Hello, World!")

root = tk.Tk()
root.geometry("200x200")

btn = tk.Button(root, text="Click Me", command=say_hello)
btn.pack()
root.mainloop()
```

In the above example, we import the tkinter module and define a function that prints "Hello, World!" when called. We then create a Tk object, set its geometry, and create a button widget that calls the say_hello function when clicked. Finally, we pack the button widget and start the main event loop.

GUI Programming with PyQt

PyQt provides a more powerful and feature-rich toolkit compared to Tkinter. Here is a simple example of how to create a GUI application using PyQt:

```
import sys
```

```
from PyQt5.QtWidgets import QApplication, QMainWindow, QPushButton

class MyWindow(QMainWindow):
    def __init__(self):
        super().__init__()

        self.setGeometry(100, 100, 300, 300)
        self.setWindowTitle("My Window")

        btn = QPushButton("Click Me", self)
        btn.setGeometry(100, 100, 100, 30)
        btn.clicked.connect(self.say_hello)

    def say_hello(self):
        print("Hello, World!")

app = QApplication(sys.argv)
window = MyWindow()
window.show()
sys.exit(app.exec_())
```

In the above example, we import the necessary modules from PyQt5 and define a custom QMainWindow class. We set its geometry and title, create a button widget that calls the say_hello function when clicked, and show the main window.

Python was created in 1989 by Guido van Rossum, and first released on February 20, 1991. Its design philosophy emphasizes code readability and simplicity. It is a high-level, interpreted language and is widely used in web development, data analysis, artificial intelligence, and more. Its popularity is due to its easy-to-learn syntax and vast collection of libraries and frameworks.

• Running programs with Python

Introduction to Python Interpreter: Understanding the Core of Python

Python is a popular, high-level programming language that is widely used in software development, data analysis, and scientific computing. Python's design philosophy emphasizes code readability, and its syntax allows programmers to express concepts in fewer lines of code than would be possible in other languages. At the core of Python's capabilities is the interpreter, which is responsible for executing Python code.

What is a Python Interpreter?

A Python interpreter is a program that reads and executes Python code. It is responsible for translating Python code into machine-readable code that can be executed by the computer's hardware. The interpreter also checks for syntax errors and runs the code line by line. The interpreter plays a vital role in the development of Python applications, as it is responsible for the execution and evaluation of the code.

How does the Python Interpreter Work?

When a Python program is executed, the Python interpreter reads the code line by line and translates it into bytecode, which is a low-level form of code that can be executed by the computer's hardware. The bytecode is then executed by the interpreter, which performs the necessary computations and returns the results to the user. The interpreter also checks for syntax errors, such as missing parentheses or invalid variable names, and reports them to the user.

Why is the Python Interpreter Important?

The Python interpreter is an essential component of the Python language, as it is responsible for executing Python code. The interpreter's ability to execute code line by line makes it easy for developers to debug their code and identify errors. Additionally, the interpreter's ability to translate Python code into bytecode allows Python to be used on a wide variety of hardware platforms, including smartphones, desktop computers, and even supercomputers.

Conclusion

In conclusion, the Python interpreter is a vital component of the Python language. It is responsible for executing Python code, checking for syntax errors, and translating Python code into bytecode. Python's design philosophy emphasizes code readability, and its syntax allows programmers to express concepts in fewer lines of code than would be possible in other languages. With its popularity in software development, data analysis, and scientific computing, Python's interpreter plays a critical role in the success of Python as a language.

Getting Started with Python Scripting Using Visual Code

Python is a high-level programming language that is widely used for general-purpose programming. It is versatile and can be used for a variety of tasks such as web development, data analysis, machine learning, and more. Python is known for its simple syntax, which makes it easy to learn and use. In this article, we will explore how to write and execute the first Python script using Visual Code.

Installing Visual Code

Visual Code is a popular open-source code editor developed by Microsoft. It is free, cross-platform, and supports many programming languages, including Python. To install Visual Code, visit the official website and download the installer for your operating system. Once downloaded, run the installer and follow the prompts to complete the installation.

Setting Up the Python Environment

Before we can start writing Python scripts in Visual Code, we need to set up the Python environment. Visual Code supports multiple Python interpreters, so you can choose the version of Python that best suits your needs. To set up the Python environment, follow these steps:

Open Visual Code and click on the Extensions tab on the left-hand side.

Search for Python in the search bar and click on Install to install the Python extension for Visual Code.

Once installed, click on the Python button on the left-hand side to open the Python Interactive window.

From the Python Interactive window, you can choose the interpreter you want to use. Click on the Select Interpreter button and choose the interpreter you want to use.

Writing the First Python Script

Now that we have set up the Python environment, we can start writing our first Python script. To create a new Python file, follow these steps:

Click on the File menu and select New File.

Type in the code for your Python script. For example, you can print the message "Hello, World!" using the following code:

```
print("Hello, World!")
```

Save the file with a .py extension. For example, you can save the file as hello.py.
Executing the Python Script

To execute the Python script in Visual Code, follow these steps:

Click on the Terminal menu and select New Terminal.

In the terminal window, navigate to the directory where your Python script is saved. For example, if your script is saved in the Documents folder, type the following command:

```
cd Documents
```

Once you are in the correct directory, type the following command to execute the Python script:
```
python hello.py
```

The message "Hello, World!" will be printed in the terminal window.

Python is a widely-used programming language that is known for its simplicity, readability, and flexibility. One of the key features of Python is its core data types, which are the building blocks of any program. In this article, we will explore Python's core data types and their functionality, without using the article word.

Introduction to Python Core Data Types

Python is a dynamically-typed language, which means that the data type of a variable is determined at runtime. This flexibility is possible because Python has several core data types, which can be used to store and manipulate different types of data.

Numeric Data Types

Python supports several numeric data types, including integers, floating-point numbers, and complex numbers. Integers are whole numbers, while floating-point numbers have decimal places. Complex numbers are numbers with a real and imaginary component.

Sequences

Sequences are ordered collections of elements. Python has three main types of sequences: lists, tuples, and ranges. Lists and tuples can contain any type of data, while ranges are used to represent a range of numbers.

Strings

Strings are sequences of characters. They can be created using single or double quotes, and can be manipulated using a variety of methods. In Python, strings are immutable, meaning that once they are created, they cannot be changed.

Sets and Dictionaries

Sets and dictionaries are used to store collections of data that are not ordered. Sets contain unique elements, while dictionaries store key-value pairs.

Converting Between Data Types

Python makes it easy to convert between different data types. For example, you can convert a string to an integer using the int() function, or convert a list to a set using the set() function.

In conclusion, Python's core data types are an essential part of the language's flexibility and ease-of-use. By understanding these data types and their functionality, you can write more efficient and effective code. Whether you are a beginner or an experienced programmer, mastering Python's core data types is a crucial step in becoming a skilled Python developer.

Setting up your Python Environment: A Comprehensive Guide

Python is a powerful programming language that has been gaining popularity over the years, thanks to its simple syntax, powerful libraries, and versatility. Before you can start programming in Python, you need to set up your Python environment. In this comprehensive guide, we will go through the steps required to set up a Python environment on your computer, without using the article word.

Downloading Python

The first step in setting up your Python environment is to download Python from the official website. Choose the appropriate version of Python for your operating system and download the installer.

Installing Python

Once you have downloaded the installer, run it and follow the installation wizard. Make sure to choose the correct options for your installation, such as adding Python to your PATH environment variable.

Choosing a Text Editor or IDE
After installing Python, you will need a text editor or integrated development environment (IDE) to write and run your Python code. There are several options available, such as PyCharm, Visual Studio Code, Sublime Text, and Atom.

Installing a Text Editor or IDE

Download and install your preferred text editor or IDE. Some text editors, such as Sublime Text and Atom, are free, while others, such as PyCharm and Visual Studio Code, have both free and paid versions.

Configuring Your Environment

After installing Python and a text editor or IDE, you will need to configure your environment. This includes setting up your PATH environment variable, installing Python packages, and configuring your text editor or IDE.

Installing Python Packages

Python has a vast collection of third-party packages that you can use to extend its functionality. You can use pip, the Python package manager, to install these packages. To install a package, open your command prompt or terminal and type "pip install package_name".

Virtual Environments

Virtual environments are a way to create isolated Python environments for your projects. They allow you to install packages and manage dependencies without affecting other projects. You can create a virtual environment using the venv module, which is included in Python 3.

Setting up a Virtual Environment

To set up a virtual environment, navigate to your project directory and run "python -m venv env_name" in your command prompt or terminal. This will create a new virtual environment with the name "env_name". Activate your virtual environment by running "source env_name/bin/activate" on macOS or Linux, or "env_name\Scripts\activate" on Windows.

In conclusion, setting up your Python environment is an essential step in starting your Python programming journey. With this guide, you now have a comprehensive understanding of the steps required to set up your Python environment on your computer. With your environment set up, you can now start writing and running Python code, installing packages, and creating virtual environments. Happy coding!

• Python basic syntax and data types

Python is a popular high-level programming language that is widely used for web development, data analysis, artificial intelligence, and more. It is known for its simple syntax, readability, and ease of use. In this article, we will explore the basics of Python syntax and data types, which are essential concepts for any beginner to learn.

Variables and Data Types
In Python, variables are used to store values that can be used throughout the program. To create a variable, you simply need to choose a name for it and assign a value to it using the "=" symbol. For example, to create a variable called "x" and assign it a value of 10, you would write:

x = 10

Python supports several data types, including:

Integers: These are whole numbers, such as 1, 2, 3, and so on.
Floats: These are numbers with decimal points, such as 1.5, 2.7, and so on.
Strings: These are sequences of characters, such as "Hello, world!" or "Python is awesome!".
Booleans: These are either True or False.
To determine the data type of a variable, you can use the type() function. For example, to determine the data type of the variable "x", you would write:

print(type(x))

This would output "int", indicating that "x" is an integer.

Basic Operators
Python supports several basic operators for performing arithmetic and logical operations. These include:

Addition: "+"
Subtraction: "-"
Multiplication: "*"
Division: "/"
Modulus: "%"
Comparison operators: "<", ">", "<=", ">=", "==", "!="
For example, to add two variables "x" and "y", you would write:

z = x + y

This would assign the sum of "x" and "y" to the variable "z". Similarly, to compare two variables "a" and "b", you would write:

```
if a > b:
print("a is greater than b")
else:
print("b is greater than a")
```

This would output either "a is greater than b" or "b is greater than a" depending on the values of "a" and "b".

Conditional Statements
Conditional statements are used to execute certain code only if a certain condition is met. In Python, conditional statements are created using the "if", "else", and "elif" keywords. For example, to check if a variable "x" is greater than 10, you would write:

```
if x > 10:
print("x is greater than 10")
else:
print("x is less than or equal to 10")
```

This would output either "x is greater than 10" or "x is less than or equal to 10" depending on the value of "x".

Loops
Loops are used to execute a block of code repeatedly. In Python, there are two main types of loops: "for" loops and "while" loops. For example, to iterate over a list of numbers and print each one, you would write:

```
numbers = [1, 2, 3, 4, 5]

for num in numbers:
print(num)
```

This would output:

```
1
2
3
4
5
```

Similarly, to execute a block of code while a certain condition is true, you would use a while loop. For example, to print all even numbers between 0 and 10, you would write:

```
i = 0

while i <= 10:
if i % 2 == 0:
print(i)
i += 1
```

This would output:

```
0
2
4
6
8
10
```

Input and Output

Input and output operations are used to communicate with the user or with other parts of the program. In Python, the "input()" function is used to get input from the user, and the "print()" function is used to display output on the screen. For example, to ask the user for their name and print a greeting, you would write:

```
name = input("What is your name? ")
print("Hello, " + name + "!")
```

This would output something like "Hello, John!" if the user entered the name "John".

In addition to printing output to the screen, you can also write data to files using Python's file handling functions. For example, to write a list of numbers to a file called "numbers.txt", you would write:

```
numbers = [1, 2, 3, 4, 5]
```

```
with open("numbers.txt", "w") as file:
for num in numbers:
file.write(str(num) + "\n")
```

This would create a file called "numbers.txt" and write the numbers 1 through 5 to it, each on a separate line.

Python syntax and data types are fundamental concepts that every beginner should understand. By mastering variables, data types, operators, conditional statements, loops, and input/output operations, you will be well on your way to writing useful and powerful Python programs. Remember to practice regularly and experiment with different types of programs to gain experience and confidence in your skills.

Variables and Data Types

Variables and data types are essential concepts for any beginner to understand when learning the Python programming language. In this article, we will explore these concepts in detail and provide examples to help you better understand how they work.

What are Variables?

Variables are essentially containers that hold values, such as numbers, text, or objects. These values can then be used throughout a program, making it easier to read, write, and manipulate data. In Python, variables are created simply by assigning a value to a name using the "=" sign. For example, to create a variable called "number" and assign it the value of 5, you would write:

number = 5

Once a variable has been created, it can be referenced and manipulated throughout the program. For example, you can add two variables together, like so:

```
a = 2
b = 3
c = a + b
```

In this example, we create two variables, "a" and "b", and then create a third variable, "c", that adds the values of "a" and "b" together.

What are Data Types?

Data types are the classification of values that variables can hold. Python supports several different data types, including integers, floats, strings, and booleans. Each data type has its own unique set of properties and operations that can be performed on it.

Integers

Integers are whole numbers, such as 1, 2, 3, and so on. In Python, integers can be represented using the "int" data type. For example, to create a variable called "age" and assign it the value of 25, you would write:

age = 25

Floats

Floats are numbers with decimal points, such as 1.5, 2.7, and so on. In Python, floats can be represented using the "float" data type. For example, to create a variable called "price" and assign it the value of 4.99, you would write:

price = 4.99

Strings

Strings are sequences of characters, such as "Hello, world!" or "Python is awesome!". In Python, strings can be represented using the "str" data type. For example, to create a variable called "name" and assign it the value of "John Doe", you would write:

name = "John Doe"

Booleans

Booleans are a type of data that can only have one of two values: True or False. In Python, booleans can be represented using the "bool" data type. For example, to create a variable called "is_active" and assign it the value of True, you would write:

is_active = True

Type Conversion

Python also provides the ability to convert between different data types. For example, you can convert a string to an integer using the "int()" function, like so:

string_number = "5"
integer_number = int(string_number)

In this example, we create a variable called "string_number" and assign it the value of "5" as a string. We then use the "int()" function to convert it to an integer and assign the result to a new variable called "integer_number".

Variables and data types are fundamental concepts in Python that you will use in every program you write. By understanding how to create variables and work with different data types, you will be able to write more complex and powerful programs. Remember to practice regularly and experiment with different types of programs to gain experience and confidence in your skills.

Basic Operators

Basic Operators are essential building blocks of programming in any language, including Python. In this article, we will explore the different types of basic operators in Python and provide examples to help you understand how they work.

Arithmetic Operators

Arithmetic operators are used to perform mathematical operations on numbers in Python. Python supports the following arithmetic operators:

Addition (+)
Subtraction (-)
Multiplication (*)
Division (/)
Modulus (%)
Exponentiation (**)
Here's an example that demonstrates the use of arithmetic operators in Python:

```
a = 5
b = 2
```

Addition
```
print(a + b) # Output: 7
```

Subtraction
```
print(a - b) # Output: 3
```

Multiplication
```
print(a * b) # Output: 10
```

Division
```
print(a / b) # Output: 2.5
```

Modulus
```
print(a % b) # Output: 1
```

Exponentiation
```
print(a ** b) # Output: 25
```

Comparison Operators

Comparison operators are used to compare two values in Python and return a Boolean value (True or False) based on the result of the comparison. Python supports the following comparison operators:

Equal to (==)
Not equal to (!=)
Greater than (>)
Less than (<)
Greater than or equal to (>=)
Less than or equal to (<=)
Here's an example that demonstrates the use of comparison operators in Python:

```
a = 5
b = 3
```

Equal to
```
print(a == b) # Output: False
```

Not equal to
```
print(a != b) # Output: True
```

Greater than
```
print(a > b) # Output: True
```

Less than
```
print(a < b) # Output: False
```

Greater than or equal to
```
print(a >= b) # Output: True
```

Less than or equal to
```
print(a <= b) # Output: False
```

Logical Operators

Logical operators are used to combine Boolean values and return a Boolean value based on the result of the combination. Python supports the following logical operators:

and
or
not

Here's an example that demonstrates the use of logical operators in Python:

```
a = True
b = False
```

and
```
print(a and b) # Output: False
```

or
```
print(a or b) # Output: True
```

not
```
print(not a) # Output: False
```

Assignment Operators

Assignment operators are used to assign values to variables in Python. They combine the assignment operator (=) with an arithmetic, comparison, or logical operator. Here's an example that demonstrates the use of assignment operators in Python:

```
a = 5
b = 2
```

Addition
```
a += b
print(a) # Output: 7
```

Subtraction
```
a -= b
print(a) # Output: 5
```

Multiplication
```
a *= b
print(a) # Output: 10
```

Division
```
a /= b
print(a) # Output: 5.0
```

Modulus
```
a %= b
print(a) # Output: 1.0
```

Exponentiation

```
a **= b
print(a) # Output: 1.0
```

Basic operators are essential building blocks of programming in Python. By understanding how to use arithmetic, comparison, logical, and assignment operators, you will be able to write more complex and powerful programs. Remember to practice regularly and experiment with different types of programs to gain experience and confidence in your skills.

Conditional Statements

Conditional Statements are used in programming to execute specific code only if certain conditions are met. Python provides a variety of conditional statements that allow programmers to create complex programs. In this article, we will explore the different types of conditional statements in Python and provide examples to help you understand how they work.

If Statement

The most basic type of conditional statement in Python is the If statement. The If statement allows you to execute a block of code only if a specific condition is true. Here's an example that demonstrates the use of an If statement in Python:

```
num = 5

if num > 0:
print("Positive number")
```

Output: Positive number

In this example, the If statement checks whether the value of "num" is greater than 0. If the condition is true, the program prints "Positive number" to the console.

If-Else Statement

The If-Else statement allows you to execute one block of code if a condition is true and another block of code if the condition is false. Here's an example that demonstrates the use of an If-Else statement in Python:

```
num = -5

if num > 0:
print("Positive number")
else:
print("Negative number")
```

Output: Negative number

In this example, the If-Else statement checks whether the value of "num" is greater than 0. If the condition is true, the program prints "Positive number" to the console. If the condition is false, the program prints "Negative number" to the console.

If-Elif-Else Statement

The If-Elif-Else statement allows you to execute different blocks of code based on multiple conditions. The Elif keyword stands for "else if" and allows you to check additional conditions after the If statement. Here's an example that demonstrates the use of an If-Elif-Else statement in Python:

```
num = 0

if num > 0:
print("Positive number")
elif num == 0:
print("Zero")
else:
print("Negative number")
```

Output: Zero

In this example, the If-Elif-Else statement checks whether the value of "num" is greater than 0, equal to 0, or less than 0. If the value of "num" is greater than 0, the program prints "Positive number" to the console. If the value of "num" is equal to 0, the program prints "Zero" to the console. If the value of "num" is less than 0, the program prints "Negative number" to the console.

Nested If Statements

Nested If statements allow you to check for multiple conditions within a single If statement. Here's an example that demonstrates the use of nested If statements in Python:

```
num = 10
```

Loops

Loops are an essential concept in programming and are used to execute a block of code repeatedly. Python provides two main types of loops: the For loop and the While loop. In this article, we will explore the different types of loops in Python and provide examples to help you understand how they work.

For Loop

The For loop is used to iterate over a sequence of elements, such as a list or a string. Here's an example that demonstrates the use of a For loop in Python:

```
fruits = ["apple", "banana", "cherry"]

for fruit in fruits:
print(fruit)
```

Output:
```
apple
banana
cherry
```

In this example, the For loop iterates over each element in the "fruits" list and prints each element to the console.

You can also use the range() function with a For loop to iterate over a range of numbers. Here's an example:

```
for i in range(1, 5):
print(i)
```

Output:
```
1
2
3
4
```

In this example, the For loop iterates over a range of numbers from 1 to 4 and prints each number to the console.

While Loop

The While loop is used to execute a block of code repeatedly as long as a condition is true. Here's an example that demonstrates the use of a While loop in Python:

```
i = 1

while i <= 5:
print(i)
i += 1
```

Output:
```
1
2
3
4
5
```

In this example, the While loop executes the block of code as long as the value of "i" is less than or equal to 5. The program prints the value of "i" to the console and increments the value of "i" by 1 in each iteration of the loop.

You can also use the break and continue statements with a While loop. The break statement is used to exit the loop prematurely, while the continue statement is used to skip over an iteration of the loop. Here's an example:

```
i = 1

while i <= 10:
if i == 5:
break
if i % 2 == 0:
i += 1
continue
print(i)
i += 1
```

Output:
```
1
3
```

In this example, the While loop executes the block of code as long as the value of "i" is less than or equal to 10. However, the break statement is used to exit the loop prematurely when the value of "i" is equal to 5. The continue statement is used to skip over the iteration of the loop when the value of "i" is even. The program prints the value of "i" to the console for odd numbers only.

Loops are an important concept in programming, and Python provides two main types of loops: the For loop and the While loop. By understanding how to use For and While loops in Python, you can create more complex programs that can iterate over a sequence of elements or execute a block of code repeatedly as long as a certain condition is true.

Input and Output

Python provides built-in functions for input and output operations, which are essential for interacting with users and reading/writing data to files. In this article, we will explore the different ways to perform input and output operations in Python with examples.

Input Operations

The input() function in Python is used to read input from the user. It prompts the user to enter input, reads the input from the console, and returns the value as a string. Here's an example that demonstrates the use of the input() function in Python:

```python
name = input("Enter your name: ")
print("Hello, " + name)
```

Output:
Enter your name: John
Hello, John

In this example, the input() function prompts the user to enter their name. The user enters their name "John", which is then stored in the variable "name". The program then uses the value of "name" to print a personalized greeting to the user.

Output Operations

The print() function in Python is used to display output to the console. It takes one or more arguments and prints them to the console. Here's an example that demonstrates the use of the print() function in Python:

```python
print("Hello, World!")
```

Output:
Hello, World!

In this example, the print() function takes a string argument "Hello, World!" and prints it to the console.

Reading from and Writing to Files

Python provides built-in functions to read from and write to files. The open() function is used to open a file, and the read() and write() functions are used to read from and write to the file, respectively. Here's an example that demonstrates the use of these functions in Python:

```
Writing to a file
with open("example.txt", "w") as f:
f.write("Hello, World!")
```

```
Reading from a file
with open("example.txt", "r") as f:
contents = f.read()
print(contents)
```

```
Output:
Hello, World!
```

In this example, the with statement is used to open the file "example.txt" in write mode and write the string "Hello, World!" to the file. The same file is then opened in read mode using the with statement, and the contents of the file are read using the read() function. The contents are then printed to the console.

Formatted Output

Python provides several ways to format output using placeholders and formatting strings. Here's an example that demonstrates the use of formatted output in Python:

```
name = "John"
age = 30
```

```
Using placeholders
print("My name is {} and I am {} years old.".format(name, age))
```

```
Using f-strings
print(f"My name is {name} and I am {age} years old.")
```

```
Output:
My name is John and I am 30 years old.
My name is John and I am 30 years old.
```

In this example, the placeholders "{}" are used to format the output string with the values of the variables "name" and "age". The same output can also be achieved using f-strings, which allow you to embed expressions inside string literals using curly braces.

Input and output operations are an essential part of programming, and Python provides built-in functions to perform these operations efficiently. By understanding how to use the input(), print(), open(), read(), and write() functions, as well as formatted output, you can create more complex programs that can read input from users, write output to files, and display formatted output to the console.

```python
if num > 0:
if num % 2 == 0:
print("Positive even number")
else:
print("Positive odd number")
else:
print("Negative number")
```

Output: Positive even number

In this example, the nested If statements first check whether the value of "num" is greater than 0. If the value of "num" is greater than 0, the program checks whether the value of "num" is even or odd. If the value of "num" is even, the program prints "Positive even number" to the console. If the value of "num" is odd, the program prints "Positive odd number" to the console. If the value of "num" is less than or equal to 0, the program prints "Negative number" to the console.

Control structures: if/else, for/while loops

Control structures are an essential part of programming that enable us to control the flow of execution in our code. In Python, the two main control structures are if/else statements and for/while loops. In this article, we will explore these control structures and demonstrate how they can be used in Python.

If/Else Statements

If/else statements allow us to execute different blocks of code based on certain conditions. In Python, if/else statements have the following syntax:

```
if condition:
# block of code to execute if condition is true
else:
# block of code to execute if condition is false
```

Here's an example that demonstrates the use of if/else statements in Python:

```
age = 25

if age >= 18:
print("You are an adult")
else:
print("You are not an adult")
```

Output:
You are an adult

In this example, the if/else statement checks if the value of "age" is greater than or equal to 18. Since the value of "age" is 25, the condition is true and the program prints "You are an adult" to the console.

For Loops

For loops are used to iterate over a sequence of values, such as a list or a tuple. In Python, for loops have the following syntax:

```
for variable in sequence:
# block of code to execute for each value in sequence
```

Here's an example that demonstrates the use of for loops in Python:

```
fruits = ["apple", "banana", "cherry"]

for fruit in fruits:
print(fruit)
```

Output:
apple
banana
cherry

In this example, the for loop iterates over the list of fruits and prints each fruit to the console.

While Loops

While loops are used to execute a block of code repeatedly as long as a certain condition is true. In Python, while loops have the following syntax:

```
while condition:
# block of code to execute while condition is true
```

Here's an example that demonstrates the use of while loops in Python:

```
i = 0

while i < 5:
print(i)
i += 1
```

Output:
0
1
2
3
4

In this example, the while loop executes the block of code as long as the value of "i" is less than 5. The value of "i" is initially 0, and it is incremented by 1 after each iteration of the loop.

Nested Control Structures

It is also possible to use nested control structures in Python, such as an if statement inside a for loop or a while loop inside an if statement. Here's an example that demonstrates a nested control structure in Python:

```
numbers = [1, 2, 3, 4, 5]

for number in numbers:
if number % 2 == 0:
print(f"{number} is even")
else:
print(f"{number} is odd")
```

Output:
1 is odd
2 is even
3 is odd
4 is even
5 is odd

In this example, the for loop iterates over the list of numbers and checks if each number is even or odd using an if statement. The program then prints a message to the console indicating whether the number is even or odd.

Control structures are an essential part of programming that enable us to control the flow of execution in our code. By understanding how to use if/else statements, for loops, and while loops, as well as nested control structures, you can create more complex programs that can perform a wide variety of tasks.

Functions and Modules: A Comprehensive Guide to Python Programming

Functions and modules are fundamental concepts in Python programming. They allow you to organize your code, make it more reusable, and improve its overall structure. In this guide, we will explore the world of functions and modules in Python, covering everything from the basics of defining and calling functions to creating and using packages.

Part 1: Introduction to Functions in Python

Functions are reusable pieces of code that perform a specific task. They can take inputs and return outputs, making them incredibly versatile. In Python, you can define a function using the "def" keyword followed by the function name, input arguments, and the code that defines what the function does. Here is an example:

```
def add_numbers(x, y):
    return x + y
```

In this example, we defined a function called "add_numbers" that takes two input arguments, "x" and "y". The function then returns the sum of "x" and "y". To call this function, we can simply pass two values to it, like this:

```
result = add_numbers(3, 5)
print(result)
```

This would output "8", which is the sum of 3 and 5.

Part 2: Working with Modules in Python

Modules are files containing Python code that can be imported into other Python code. They allow you to reuse code across multiple programs and make your code more organized. Python comes with many built-in modules, such as "math" and "random", which provide functions for mathematical calculations and generating random numbers, respectively.

To use a module in your code, you need to import it first. You can do this using the "import" keyword, followed by the name of the module. Here is an example:

```
import math

result = math.sqrt(25)
print(result)
```

In this example, we imported the "math" module and used its "sqrt" function to calculate the square root of 25. This would output "5.0".

Part 3: Advanced Function Concepts in Python

Python functions can be quite powerful, and there are many advanced concepts that you can use to make them even more versatile. Some of these concepts include default arguments, variable-length argument lists, and recursion.

Default arguments allow you to define a default value for an input argument, which is used if no value is provided when the function is called. Here is an example:

```
def greet(name="World"):
    print("Hello, " + name + "!")

greet("Alice")
greet()
```

In this example, we defined a function called "greet" that takes a default argument of "World" for the "name" input. When we call the function with the argument "Alice", it outputs "Hello, Alice!". When we call the function without an argument, it uses the default value and outputs "Hello, World!".

Variable-length argument lists allow you to define a function that can take a variable number of input arguments. Here is an example:

```
def sum_numbers(*args):
    result = 0
    for num in args:
        result += num
    return result

print(sum_numbers(1, 2, 3))
print(sum_numbers(4, 5, 6, 7))
```

In this example, we defined a function called "sum_numbers" that takes a variable-length argument list using the "*" operator. The function then iterates over the input arguments and returns their sum. When we call the function with the arguments 1, 2, and 3, it outputs "6". When we call the function with the arguments 4, 5, 6, and 7, it outputs "22".

Recursion is a technique where a function calls itself. This can be useful for solving problems that can be broken down into smaller subproblems. Here is an example:

```
def factorial(n):
    if n == 0:
        return 1
    else:
        return n * factorial(n-1)

print(factorial(5))
```

In this example, we defined a function called "factorial" that calculates the factorial of a number using recursion. The base case of the recursion is when "n" is equal to 0, in which case the function returns 1. Otherwise, the function multiplies "n" by the result of calling "factorial" with "n-1". When we call the function with the argument 5, it outputs "120", which is the factorial of 5.

Part 4: Creating and Using Packages in Python

Packages are collections of modules that can be used to organize your code and create reusable libraries. To create a package in Python, you need to create a directory with a special file called "init.py". This file tells Python that the directory is a package and can contain other modules.

Here is an example of how to create a simple package:

Create a new directory called "my_package".
Create a file called "init.py" inside the "my_package" directory.
Create a file called "my_module.py" inside the "my_package" directory.
Add some code to the "my_module.py" file, such as a function that adds two numbers.
Import the package and module in your Python code and use the function.
Here is an example:

```
# Inside my_module.py
def add_numbers(x, y):
    return x + y

# Inside your Python code
import my_package.my_module

result = my_package.my_module.add_numbers(3, 5)
print(result)
```

In this example, we created a package called "my_package" that contains a module called "my_module". The "my_module" module contains a function called "add_numbers" that adds two numbers. We then imported the package and module into our Python code and used the "add_numbers" function to add 3 and 5, which outputs "8".

Part 5: Best Practices for Functions and Modules in Python

When writing functions and modules in Python, it is important to follow best practices to ensure that your code is easy to read, maintain, and reuse. Some best practices include:

Writing modular and reusable code
Using descriptive variable and function names
Writing docstrings for your code
Using proper indentation and whitespace
Following naming conventions, such as using lowercase names for functions and modules
By following these best practices, you can create high-quality Python code that is easy to understand and use.

Functions and modules are essential concepts in Python programming. They allow you to organize your code, make it more reusable, and improve its overall structure. By understanding the basics of defining and calling functions, importing and using modules, and using advanced function concepts, you can create powerful Python programs. Additionally, by creating and using packages and following best practices for functions and modules, you can make your code more organized, maintainable, and reusable.

Input/Output in Python

Part 1: Introduction to Input/Output in Python

Input and output (I/O) are important concepts in programming. In Python, you can use various functions to read and write data from and to different sources, such as files, the console, or the network. I/O operations are critical for building applications that interact with users, data storage, or other systems.

Part 2: Reading Data from Files in Python

Reading data from files is a common task in Python. You can read data from various file formats, such as text, CSV, JSON, or binary files. Python provides several built-in functions to read data from files, such as "open()" and "read()".

Here is an example of how to read data from a text file:

```
file = open("data.txt", "r")
content = file.read()
file.close()

print(content)
```

In this example, we first open a file called "data.txt" in read mode ("r"). We then read the content of the file using the "read()" function and store it in a variable called "content". Finally, we close the file using the "close()" function and print the content.

Part 3: Writing Data to Files in Python

Writing data to files is also a common task in Python. You can write data to various file formats, such as text, CSV, JSON, or binary files. Python provides several built-in functions to write data to files, such as "open()" and "write()".

Here is an example of how to write data to a text file:

```
file = open("data.txt", "w")
file.write("Hello, world!")
file.close()
```

In this example, we first open a file called "data.txt" in write mode ("w"). We then write the string "Hello, world!" to the file using the "write()" function. Finally, we close the file using the "close()" function.

Part 4: Reading and Writing CSV Files in Python

CSV (comma-separated values) files are a common file format for storing and exchanging tabular data. In Python, you can use the "csv" module to read and write CSV files. This module provides various functions and classes to handle CSV files, such as "csv.reader()", "csv.writer()", and "csv.DictReader()".

Here is an example of how to read a CSV file using the "csv" module:

```
import csv

with open("data.csv", "r") as file:
    reader = csv.reader(file)
    for row in reader:
        print(row)
```

In this example, we first import the "csv" module. We then open a file called "data.csv" in read mode using a "with" statement, which automatically closes the file after we're done reading it. We then create a reader object using the "csv.reader()" function and loop over the rows in the file using a "for" loop. Finally, we print each row to the console.

Here is an example of how to write a CSV file using the "csv" module:

```
import csv

data = [
    ["Name", "Age", "City"],
    ["John", "25", "New York"],
    ["Jane", "30", "San Francisco"],
    ["Bob", "35", "Chicago"],
]

with open("data.csv", "w", newline="") as file:
    writer = csv.writer(file)
    writer.writerows(data)
```

In this example, we first import the "csv" module. We then define a list of lists called "data", where each inner list represents a row in the CSV file. We then open a file called "data.csv" in write mode using a "with" statement and specify the "newline" parameter to prevent the writer from inserting additional line breaks. We then create a writer object using the "csv.writer()" function and use the "writerows()" method to write the data to the file.

Part 5: Reading and Writing JSON Files in Python

JSON (JavaScript Object Notation) is a lightweight data interchange format that is easy for humans to read and write and easy for machines to parse and generate. In Python, you can use the "json" module to read and write JSON files. This module provides various functions and classes to handle JSON files, such as "json.load()", "json.loads()", "json.dump()", and "json.dumps()".

Here is an example of how to read a JSON file using the "json" module:

```
import json

with open("data.json", "r") as file:
    data = json.load(file)

print(data)
```

In this example, we first import the "json" module. We then open a file called "data.json" in read mode using a "with" statement, which automatically closes the file after we're done reading it. We then use the "json.load()" function to parse the JSON data from the file and store it in a variable called "data". Finally, we print the "data" variable to the console.

Here is an example of how to write a JSON file using the "json" module:

```
import json

data = {
    "name": "John",
    "age": 25,
    "city": "New York"
}

with open("data.json", "w") as file:
    json.dump(data, file)
```

In this example, we first import the "json" module. We then define a dictionary called "data" that contains some data to be written to the JSON file. We then open a file called "data.json" in write mode using a "with" statement and create a JSON encoder using the "json.dump()" function. We then use the "dump()" method to encode the data as JSON and write it to the file.

Part 6: Reading and Writing Binary Files in Python

Binary files are files that contain data in a binary format, which is different from the text-based formats we've seen so far. Binary files can contain any type of data, such as images, audio, video, or serialized objects. In Python, you can use the "open()" function with the "rb" and "wb" modes to read and write binary files.

Here is an example of how to read a binary file in Python:

```
with open("data.bin", "rb") as file:
    data = file.read()

print(data)
```

In this example, we open a file called "data.bin" in read mode with the "rb" mode. We then read the contents of the file using the "read()" method and store it in a variable called "data". Finally, we print the "data" variable to the console.

Here is an example of how to write a binary file in Python:

```
data = b'\x00\x01\x02\x03\x04'

with open("data.bin", "wb") as file:
    file.write(data)
```

In this example, we define a bytes object called "data" that contains some binary data to be written to the file. We then open a file called "data.bin" in write mode with the "wb" mode and write the "data" object to the file using the "write()" method.

Working with files

Part 1: Introduction

Working with files is an essential part of programming, and Python provides various functions and modules to handle file operations. In this article, we'll explore the different ways of working with files in Python and see some code examples.

Part 2: Opening and Closing Files in Python

Before we can work with a file in Python, we need to open it using the "open()" function. The "open()" function takes two arguments: the filename and the mode in which to open the file. The mode can be "r" for reading, "w" for writing, "a" for appending, "x" for exclusive creation, "b" for binary mode, and "t" for text mode.

Here is an example of how to open a file for reading:

```
file = open("data.txt", "r")
```

In this example, we open a file called "data.txt" in read mode and store the file object in a variable called "file". Once we're done working with the file, we need to close it using the "close()" method:

```
file.close()
```

This method ensures that any changes made to the file are saved, and any resources used by the file are freed.

Part 3: Reading Files in Python

Once we've opened a file for reading, we can read its contents using various methods provided by Python. Here is an example of how to read the entire contents of a file:

```
with open("data.txt", "r") as file:
    data = file.read()

print(data)
```

In this example, we open a file called "data.txt" in read mode using a "with" statement, which automatically closes the file after we're done reading it. We then use the "read()" method to read the entire contents of the file and store it in a variable called "data". Finally, we print the "data" variable to the console.

Here is an example of how to read a file line by line:

```
with open("data.txt", "r") as file:
    for line in file:
        print(line)
```

In this example, we open a file called "data.txt" in read mode using a "with" statement, which automatically closes the file after we're done reading it. We then use a "for" loop to iterate over the file object and print each line to the console.

Part 4: Writing Files in Python

Once we've opened a file for writing, we can write data to it using various methods provided by Python. Here is an example of how to write a string to a file:

```
with open("data.txt", "w") as file:
    file.write("Hello, world!")
```

In this example, we open a file called "data.txt" in write mode using a "with" statement, which automatically closes the file after we're done writing to it. We then use the "write()" method to write the string "Hello, world!" to the file.

Here is an example of how to write a list of strings to a file:

```
data = ["apple", "banana", "orange"]

with open("data.txt", "w") as file:
    for item in data:
        file.write(item + "\n")
```

In this example, we define a list of strings called "data" that we want to write to a file. We then open a file called "data.txt" in write mode using a "with" statement, which automatically closes the file after we're done writing to it. We then use a "for" loop to iterate over the list of strings and write each item to the file,separating them by a new line character ("\n").

Part 5: Appending to Files in Python

If we want to add new data to an existing file, we can open the file in "append" mode using the "a" argument. Here is an example of how to append a string to a file:

```
with open("data.txt", "a") as file:
    file.write("This is a new line!")
```

In this example, we open a file called "data.txt" in append mode using a "with" statement, which automatically closes the file after we're done appending to it. We then use the "write()" method to append the string "This is a new line!" to the file.

Part 6: Working with Binary Files in Python

In addition to working with text files, Python can also handle binary files such as images, videos, and audio files. Here is an example of how to read the contents of a binary file:

```
with open("image.jpg", "rb") as file:
    data = file.read()

print(data)
```

In this example, we open a file called "image.jpg" in binary mode using a "with" statement, which automatically closes the file after we're done reading it. We then use the "read()" method to read the entire contents of the file and store it in a variable called "data". Finally, we print the "data" variable to the console.

Part 7: Working with Files in Different Directories

By default, Python looks for files in the same directory as the script that's running. However, we can also specify the path to a file to work with files in different directories. Here is an example of how to open a file in a different directory:

```
with open("/path/to/file/data.txt", "r") as file:
    data = file.read()

print(data)
```

In this example, we specify the path to the file as "/path/to/file/data.txt" and open it in read mode using a "with" statement. We then use the "read()" method to read the entire contents of the file and store it in a variable called "data". Finally, we print the "data" variable to the console.

we explored the different ways of working with files in Python, including opening and closing files, reading and writing data, and working with binary files. We also saw some code examples that illustrate how to work with files in Python. With these concepts and techniques, you can easily handle file operations in your Python programs.

Working with files

Python is a popular programming language used for various purposes such as web development, data analysis, and machine learning. One of the essential aspects of programming is working with files. Files are used to store and retrieve data, and Python provides numerous ways to interact with them. In this article, we will explore various methods to work with files in Python.

Opening and Closing Files:
Before we can work with a file, we need to open it. In Python, we can use the open() function to open a file. The function takes two parameters: the path to the file and the mode in which we want to open it. Modes can be 'r' for reading, 'w' for writing, and 'a' for appending to an existing file. Once we are done with the file, we need to close it using the close() function. Failing to close a file can result in memory leaks and data corruption.

Reading Files:

Python provides several ways to read data from a file. One of the most common methods is to read the entire file using the read() function. The function returns a string containing the contents of the file. Another way is to read the file line by line using the readline() function. The function reads a single line from the file and returns it as a string. We can also use the readlines() function to read all the lines of the file and return them as a list of strings.

Writing to Files:
To write data to a file, we need to open the file in 'w' or 'a' mode. 'w' mode overwrites the existing file, while 'a' mode appends to an existing file. We can write to a file using the write() function, which takes a string as a parameter. We can also write multiple lines of data using the writelines() function, which takes a list of strings as a parameter.

Closing Files:

As mentioned earlier, we need to close a file after we are done working with it. In Python, we can use the with statement to automatically close a file after we are done with it. The with statement ensures that the file is closed even if an error occurs. Here's an example:

```
with open('example.txt', 'r') as file:
    # Perform operations on the file
```

Working with Binary Files:
In addition to text files, we can also work with binary files in Python. Binary files contain non-textual data such as images, audio, and video. To read and write binary files, we need to open them in binary mode using the 'b' character in the mode parameter. Here's an example of opening a binary file:

```
with open('example.bin', 'rb') as file:
    # Perform operations on the binary file
```

Working with files is an essential aspect of programming, and Python provides various methods to interact with them. We can read and write text files, as well as binary files. It's crucial to remember to close files after we are done working with them to avoid memory leaks and data corruption. The with statement can be used to automatically close files and ensure that they are closed even if an error occurs.

Error handling in Python

Error handling is a critical aspect of programming. As programmers, we can never assume that everything will always go as planned. Python provides several ways to handle errors in our code. In this article, we will explore different error handling techniques in Python.

What is an Error?

In programming, an error is a deviation from the expected behavior of a program. Errors can be classified into two types: syntax errors and runtime errors. Syntax errors occur when the code violates the rules of the language. Runtime errors occur when the code is syntactically correct but encounters an issue during execution.

Types of Errors:

There are several types of errors in Python, such as NameError, TypeError, ValueError, ZeroDivisionError, and ImportError. Each error type indicates a specific problem in the code. For example, NameError occurs when we try to use a variable that has not been defined, while TypeError occurs when we try to perform an operation on incompatible types.

Try and Except Blocks:
Python provides a built-in exception handling mechanism using the try and except blocks. We can use the try block to wrap the code that might raise an exception. If an exception occurs, the except block is executed. Here's an example:

```
try:
    # Code that might raise an exception
except ExceptionType:
    # Code to handle the exception
```

In this example, ExceptionType is the type of exception we want to handle. We can also use a bare except block to catch all types of exceptions. However, it's generally not a good practice as it can mask other issues in the code.

Raising Exceptions:

We can also raise our own exceptions using the raise statement. We can raise exceptions for specific conditions in our code. Here's an example:

```
if x < 0:
    raise ValueError("x cannot be negative")
```

In this example, we raise a ValueError if x is negative. We can also create our own custom exception classes by subclassing the built-in Exception class.

Finally Block:
In addition to the try and except blocks, Python also provides a finally block. The code in the finally block is always executed, regardless of whether an exception occurs or not. Here's an example:

```
try:
    # Code that might raise an exception
except ExceptionType:
    # Code to handle the exception
finally:
    # Code that is always executed
```

Error handling is a crucial aspect of programming, and Python provides several ways to handle errors in our code. We can use the try and except blocks to catch and handle exceptions, raise our own exceptions using the raise statement, and use the finally block to ensure that code is always executed, regardless of exceptions. By using these error handling techniques, we can make our code more robust and reliable.

Object-Oriented Programming (OOP) in Python

Object-Oriented Programming (OOP) is a programming paradigm that focuses on creating objects that encapsulate data and behavior. OOP enables programmers to write code that is more modular, extensible, and reusable. Python is an object-oriented language that supports OOP principles. In this article, we will explore the fundamentals of OOP in Python.

Classes and Objects:
In Python, a class is a blueprint for creating objects. It defines the attributes and methods that an object of that class will have. An object is an instance of a class. We can create multiple objects from the same class, each with its own unique data.

```
class MyClass:
    def __init__(self, arg1, arg2):
        self.arg1 = arg1
        self.arg2 = arg2

obj1 = MyClass(1, 2)
obj2 = MyClass(3, 4)
```

In this example, we define a MyClass class with two attributes arg1 and arg2. We then create two objects obj1 and obj2 from the MyClass class, passing in different arguments to their constructors.

Inheritance:

Inheritance is a mechanism by which one class can inherit attributes and methods from another class. The class that inherits from another class is called a subclass or derived class, and the class that is inherited from is called the superclass or base class.

```
class MySubClass(MyClass):
    def my_method(self):
        # Code that uses inherited attributes and methods
```

In this example, we define a MySubClass subclass that inherits from the MyClass superclass. The MySubClass subclass can use the attributes and methods of the MyClass superclass, in addition to defining its own unique attributes and methods.

Polymorphism:

Polymorphism is the ability of objects to take on different forms. In Python, polymorphism can be achieved through method overriding and method overloading. Method overriding is when a subclass redefines a method that is already defined in the superclass. Method overloading is when a class has multiple methods with the same name but different arguments.

```
class MySubClass(MyClass):
    def my_method(self):
        # Override the superclass method

    def my_method(self, arg1):
        # Overload the method with a different number of arguments
```

In this example, we define a MySubClass subclass that overrides the my_method method defined in the MyClass superclass. We also define an overloaded version of my_method that takes in an additional argument.

Encapsulation:

Encapsulation is the practice of hiding the internal details of an object from the outside world. In Python, encapsulation can be achieved through the use of private attributes and methods.

```
class MyClass:
    def __init__(self):
        self._my_private_attr = 1

    def my_public_method(self):
        # Code that uses the private attribute
```

In this example, we define a MyClass class with a private attribute _my_private_attr. The attribute is marked as private by prefixing it with an underscore. The class also has a public method my_public_method that can access the private attribute.

Object-Oriented Programming is a powerful paradigm that can make our code more modular, extensible, and reusable. In Python, we can use classes and objects to encapsulate data and behavior, inherit attributes and methods from other classes, achieve polymorphism through method overriding and overloading, and encapsulate our objects' internal details using private attributes and methods. By using these OOP principles, we can write code that is easier to maintain and more robust.

Inheritance and polymorphism

Inheritance and polymorphism are two core concepts of Object-Oriented Programming (OOP) that allow us to write more efficient and reusable code. In Python, these concepts are fundamental and powerful tools that allow developers to create complex software with ease. In this article, we will explore the basics of inheritance and polymorphism in Python and provide some examples to illustrate their usage.

Inheritance:

Inheritance is a mechanism in OOP that allows us to create a new class by deriving properties and characteristics from an existing class. The existing class is called the base class or the superclass, and the new class is called the derived class or the subclass.

```python
class Animal:
    def __init__(self, name, age):
        self.name = name
        self.age = age

class Dog(Animal):
    def bark(self):
        print("Woof!")
```

In this example, we have defined two classes: Animal and Dog. The Dog class is a subclass of the Animal class, and it inherits the name and age attributes from the Animal class. The Dog class also has a unique method called bark, which is not present in the Animal class.

Polymorphism:

Polymorphism is another essential concept in OOP, which means that objects of different classes can be treated as if they were objects of the same class. It allows us to write more generic and flexible code that can work with different types of objects.class Animal:

```python
class Animal:
    def __init__(self, name, age):
        self.name = name
        self.age = age

    def make_sound(self):
        pass

class Dog(Animal):
    def make_sound(self):
```

```
    print("Woof!")

class Cat(Animal):
    def make_sound(self):
        print("Meow!")
```

In this example, we have defined three classes: Animal, Dog, and Cat. All three classes have a method called make_sound, but the implementation of the method is different in each class. The Animal class has a blank implementation of the make_sound method, while the Dog and Cat classes override the method and implement their own unique behavior.

Inheritance and Polymorphism Combined:

Inheritance and polymorphism can be combined to create more complex and powerful software systems. The derived classes can inherit properties and methods from their base classes, and they can also override and implement their own unique behavior.

```
class Vehicle:
    def __init__(self, name, model):
        self.name = name
        self.model = model

    def drive(self):
        print("Driving...")

class Car(Vehicle):
    def __init__(self, name, model, num_doors):
        super().__init__(name, model)
        self.num_doors = num_doors

    def drive(self):
        print("Driving car...")

class Truck(Vehicle):
    def __init__(self, name, model, payload_capacity):
        super().__init__(name, model)
        self.payload_capacity = payload_capacity

    def drive(self):
        print("Driving truck...")

vehicles = [Car("Honda", "Civic", 4), Truck("Ford", "F-150", 2000)]

for vehicle in vehicles:
    vehicle.drive()
```

In this example, we have defined a Vehicle class that has a drive method. We then defined two derived classes: Car and Truck, which inherit from the Vehicle class and override the drive method with their unique implementations. We create a list of Car and Truck objects and call the drive method on each object.

Regular expressions in Python

Regular expressions are a powerful tool for pattern matching and text manipulation in Python. Whether you're parsing a large dataset or searching for specific strings in a document, regular expressions can help you get the job done quickly and efficiently.

What are Regular Expressions?

At their core, regular expressions are a way to describe patterns in text. They allow you to search for specific sequences of characters, such as all words that start with "cat" or all phone numbers formatted in a certain way.

Regular expressions use a combination of metacharacters and literal characters to define the patterns you're searching for. For example, the "." metacharacter matches any single character, while the "*" metacharacter matches zero or more occurrences of the preceding character.

Using Regular Expressions in Python

Python provides a built-in module for regular expressions called "re". This module provides a number of functions for working with regular expressions, including searching for patterns, replacing patterns, and splitting text based on patterns.

To use regular expressions in Python, you first need to import the "re" module:

```
import re
```

Searching for Patterns

One of the most common uses of regular expressions is searching for patterns in text. The "re.search()" function allows you to search for a pattern in a string and return the first match:

Searching for Patterns

One of the most common uses of regular expressions is searching for patterns in text. The "re.search()" function allows you to search for a pattern in a string and return the first match:

In this example, we're searching for the word "brown" in the string "The quick brown fox jumps over the lazy dog". The "re.search()" function returns a match object, which we can use to get the matched text using the "group()" method.

Replacing Patterns

Another useful feature of regular expressions is the ability to replace patterns in text. The "re.sub()" function allows you to search for a pattern and replace it with a new string:

```
text = "The quick brown fox jumps over the lazy dog"
pattern = "brown"

new_text = re.sub(pattern, "red", text)
print(new_text)
```

In this example, we're searching for the word "brown" and replacing it with the word "red". The "re.sub()" function returns a new string with the pattern replaced.

Splitting Text

Regular expressions can also be used to split text based on patterns. The "re.split()" function allows you to split a string into a list of substrings based on a pattern:

```
text = "The quick brown fox jumps over the lazy dog"
pattern = "\s+" # split on one or more whitespace characters

substrings = re.split(pattern, text)
print(substrings)
```

In this example, we're splitting the string based on one or more whitespace characters. The "re.split()" function returns a list of substrings.

Regular expressions are a powerful tool for pattern matching and text manipulation in Python. They allow you to search for specific patterns in text, replace patterns with new strings, and split text into substrings based on patterns. By learning how to use regular expressions in Python, you can greatly enhance your text processing capabilities.

Lambda functions and functional programming

Lambda functions and functional programming are powerful tools in Python for creating concise and efficient code. With lambda functions, you can write simple functions on the fly without defining them beforehand. In this article, we will explore the concept of lambda functions and how they are used in functional programming.

Introduction to Lambda Functions

Lambda functions, also known as anonymous functions, are functions that are not bound to a name. They are defined using the lambda keyword and can take any number of arguments. Lambda functions are commonly used in functional programming to create short and simple functions.

Here is an example of a lambda function:

```
lambda x, y: x + y
```

This lambda function takes two arguments, x and y, and returns their sum. The function can be assigned to a variable and used like any other function:

```
add = lambda x, y: x + y
result = add(3, 4)
print(result) # Output: 7
```

Using Lambda Functions in Functional Programming

Functional programming is a programming paradigm that emphasizes the use of functions to create reusable code. Lambda functions are a key tool in functional programming because they allow you to create small, reusable functions on the fly.

In functional programming, lambda functions are often used with higher-order functions, which are functions that take other functions as arguments. For example, the built-in map() function in Python takes a function and applies it to every element in a sequence. Here is an example of using a lambda function with the map() function:

```
numbers = [1, 2, 3, 4, 5]
squares = map(lambda x: x ** 2, numbers)
print(list(squares)) # Output: [1, 4, 9, 16, 25]
```

In this example, we define a lambda function that takes a number and returns its square. We then use the map() function to apply the lambda function to every element in the numbers list.

Lambda functions can also be used with filter() function in Python, which filters elements from a sequence based on a condition:

```
numbers = [1, 2, 3, 4, 5, 6, 7, 8, 9, 10]
even_numbers = filter(lambda x: x % 2 == 0, numbers)
print(list(even_numbers)) # Output: [2, 4, 6, 8, 10]

numbers = [1, 2, 3, 4, 5, 6, 7, 8, 9, 10]
even_numbers = filter(lambda x: x % 2 == 0, numbers)
print(list(even_numbers)) # Output: [2, 4, 6, 8, 10]
```

In this example, we define a lambda function that takes a number and returns True if it's even. We then use the filter() function to apply the lambda function to every element in the numbers list and return only the even numbers.

Advantages of Lambda Functions

Lambda functions offer several advantages over traditional functions. First, they are concise and easy to read, which makes them ideal for simple, one-time functions. Second, they can be used in functional programming to create reusable code. Finally, lambda functions are often used in conjunction with higher-order functions, which can lead to more efficient and readable code.

Lambda functions are a powerful tool in Python for creating concise and efficient code. They are commonly used in functional programming to create small, reusable functions on the fly. By learning how to use lambda functions, you can greatly enhance your programming skills and create more efficient and readable code.

Decorators

Decorators are a powerful feature in Python that allow you to modify the behavior of a function or class without modifying its source code. They provide a clean and efficient way to add functionality to your code and make it more readable and maintainable.

Introduction to Decorators

Decorators are functions that take another function as input and return a new function as output. They are used to modify the behavior of the input function, without changing its original code. Decorators are commonly used to add functionality such as caching, logging, authentication, and performance monitoring to a function or class.

Here is an example of a decorator that adds logging functionality to a function:

```python
def log_decorator(func):
    def wrapper(*args, **kwargs):
        print(f"Calling function {func.__name__}")
        result = func(*args, **kwargs)
        print(f"Finished calling function {func.__name__}")
        return result
    return wrapper

@log_decorator
def add(x, y):
    return x + y
```

In this example, the log_decorator function is a decorator that takes a function as input and returns a new function that adds logging functionality to the original function. The wrapper function is the new function that is returned by the decorator. It takes any number of positional and keyword arguments and passes them to the original function. The result variable stores the return value of the original function, which is then returned by the wrapper function.

Using Decorators in Python

Decorators are used in Python to add functionality to functions or classes. You can use them to add features such as caching, logging, performance monitoring, and authentication to your code. Here is an example of using a decorator to add caching functionality to a function:

```python
def cache_decorator(func):
    cache = {}
    def wrapper(*args):
        if args in cache:
```

```
        return cache[args]
    result = func(*args)
    cache[args] = result
    return result
return wrapper

@cache_decorator
def fib(n):
    if n <= 1:
        return n
    return fib(n-1) + fib(n-2)
```

In this example, the cache_decorator function is a decorator that takes a function as input and returns a new function that adds caching functionality to the original function. The cache variable stores a cache of previously computed results. The wrapper function first checks if the result for the given arguments is already in the cache. If it is, it returns the cached result. Otherwise, it computes the result using the original function and stores it in the cache.

Benefits of Decorators

Decorators offer several benefits in Python. First, they allow you to add functionality to a function or class without modifying its original code, which makes your code more maintainable and reusable. Second, they allow you to separate concerns and keep your code organized by grouping related functionality together in decorators. Finally, they allow you to create clean and efficient code by eliminating the need for redundant code.

Decorators are a powerful feature in Python that allow you to modify the behavior of a function or class without changing its original code. They provide a clean and efficient way to add functionality to your code and make it more readable and maintainable. By learning how to use decorators, you can greatly enhance your programming skills and create more efficient and reusable code.

Generators and iterators

Generators and iterators are two important concepts in Python that allow you to work with large amounts of data efficiently. In this article, we'll explore how generators and iterators work, and how you can use them to write more efficient and readable code.

Introduction to Generators and Iterators

Iterators are objects that allow you to traverse a sequence of data, one element at a time. In Python, iterators are implemented as classes that define two methods: __iter__() and __next__(). The __iter__() method returns the iterator object itself, and the __next__() method returns the next element in the sequence.

Generators are a type of iterator that are defined using a special syntax. Instead of defining a class that implements the __iter__() and __next__() methods, you define a function that uses the yield keyword to return a sequence of values one at a time.

Here's an example of a generator function that generates a sequence of Fibonacci numbers:

```
def fibonacci():
    a, b = 0, 1
    while True:
        yield a
        a, b = b, a + b
```

In this example, the fibonacci() function is a generator that yields a sequence of Fibonacci numbers. Each time the yield keyword is encountered, the function returns the current value of a and then suspends execution until the next value is requested.

Using Generators and Iterators in Python

Generators and iterators are used extensively in Python to work with large datasets efficiently. They allow you to process data one element at a time, which can save memory and improve performance.

Here's an example of how to use a generator to process a large CSV file:

```
import csv

def read_csv(filename):
    with open(filename, 'r') as f:
        reader = csv.reader(f)
        next(reader) # skip header
```

```
for row in reader:
    yield row
```

In this example, the read_csv() function is a generator that reads a CSV file one row at a time. The csv.reader() function returns an iterator that yields each row as a list of strings. The next() function is used to skip the header row, and the yield keyword is used to return each row as it is read from the file.

Benefits of Generators and Iterators

Generators and iterators offer several benefits in Python. First, they allow you to work with large datasets efficiently by processing data one element at a time. This can save memory and improve performance, especially when working with datasets that are too large to fit in memory.

Second, they allow you to write more readable and maintainable code by separating the logic for processing data from the logic for iterating over it. This can make your code easier to understand and modify, especially when working with complex data structures.

Finally, they allow you to create reusable code by encapsulating the logic for processing data in a generator or iterator. This can make your code more modular and easier to reuse in other projects.

Generators and iterators are two important concepts in Python that allow you to work with large amounts of data efficiently. They offer several benefits over traditional data processing techniques, including improved performance, readability, and modularity. By learning how to use generators and iterators, you can greatly enhance your programming skills and create more efficient and reusable code.

List comprehensions

List comprehensions are a concise and efficient way to create lists in Python. They allow you to create a new list by applying an expression to each element of an existing list, while also providing the ability to filter elements based on a condition. In this article, we'll explore the basics of list comprehensions and show you how to use them in your Python code.

Understanding List Comprehensions

List comprehensions are a compact way to create a new list by applying a transformation to each element of an existing list. The basic syntax for a list comprehension is:

[expression for item in iterable]

In this syntax, expression is the transformation to apply to each element of iterable, and item is the name of the variable that represents each element of iterable. iterable is any object that can be iterated over, such as a list, tuple, or string.

Here's an example of a simple list comprehension that squares each element of a list:

```
numbers = [1, 2, 3, 4, 5]
squares = [x ** 2 for x in numbers]
print(squares) # [1, 4, 9, 16, 25]
```

In this example, the list comprehension [x ** 2 for x in numbers] applies the transformation x ** 2 to each element x of the list numbers, resulting in a new list of squared numbers.

Filtering with List Comprehensions

In addition to transforming each element of an existing list, list comprehensions also provide the ability to filter elements based on a condition. The syntax for a filtered list comprehension is:

[expression for item in iterable if condition]

In this syntax, condition is a boolean expression that is evaluated for each element of iterable. Only elements for which condition evaluates to True are included in the new list.

Here's an example of a filtered list comprehension that only includes even numbers from a list:

```
numbers = [1, 2, 3, 4, 5]
evens = [x for x in numbers if x % 2 == 0]
print(evens) # [2, 4]
```

In this example, the filtered list comprehension [x for x in numbers if x % 2 == 0] includes only those elements x of the list numbers for which x % 2 == 0 is True, resulting in a new list of even numbers.

Nested List Comprehensions

List comprehensions can also be nested, allowing you to create more complex transformations and filters. The syntax for a nested list comprehension is:

[expression for item in iterable1 if condition1 for item2 in iterable2 if condition2]

In this syntax, condition1 is a boolean expression that is evaluated for each element of iterable1. Only elements for which condition1 evaluates to True are included in the new list. item2 and iterable2 are a nested item and iterable, respectively, that are evaluated for each element that passes condition1. condition2 is a boolean expression that is evaluated for each element of iterable2.

Here's an example of a nested list comprehension that generates all pairs of numbers from two lists:

```
numbers = [1, 2, 3]
letters = ['a', 'b', 'c']
pairs = [(x, y) for x in numbers for y in letters]
print(pairs) # [(1, 'a'), (1, 'b'), (1, 'c'), (2, 'a'), (2, 'b'), (2, 'c'), (3, 'a'), (3, 'b'), (3, 'c')]
```

In this example, the nested list comprehension [(x, y) for x in numbers for y in letters] generates all pairs of numbers and letters, resulting in a list of tuples.

List comprehensions are a powerful and concise way to create lists in Python. They allow you to apply a transformation to each element of an existing list, as well as filter elements based on a condition. They can also be nested to create more complex transformations and filters. By using list comprehensions in your code, you can write more efficient and expressive Python programs.

Sets and dictionaries

Sets and dictionaries are two important data structures in Python that can be used to store and manipulate collections of data. While both sets and dictionaries are similar in some ways, they have different properties and use cases. In this article, we'll explore the basics of sets and dictionaries in Python, and show you how to use them in your code.

Understanding Sets

A set is an unordered collection of unique elements. In Python, sets are created using curly braces {} or the set() function. Here's an example of creating a set of integers:

```
numbers = {1, 2, 3, 4, 5}
```

In this example, numbers is a set containing the integers 1, 2, 3, 4, and 5.

Adding and Removing Elements from Sets

Sets provide methods for adding and removing elements. The add() method adds a single element to a set, while the update() method can be used to add multiple elements. Here's an example of using these methods:

```
numbers = {1, 2, 3}
numbers.add(4)
print(numbers) # {1, 2, 3, 4}
numbers.update([5, 6])
print(numbers) # {1, 2, 3, 4, 5, 6}
```

In this example, we first add the integer 4 to the numbers set using the add() method. We then add the integers 5 and 6 to the numbers set using the update() method.

To remove an element from a set, you can use the remove() or discard() methods. The difference between these methods is that remove() will raise a KeyError if the element is not found in the set, while discard() will simply do nothing. Here's an example:

```
numbers = {1, 2, 3, 4, 5}
numbers.remove(4)
print(numbers) # {1, 2, 3, 5}
numbers.discard(6)
print(numbers) # {1, 2, 3, 5}
```

In this example, we first remove the integer 4 from the numbers set using the remove() method. We then attempt to remove the integer 6, which is not in the set, using the discard() method. Since 6 is not in the set, nothing happens.

Understanding Dictionaries

A dictionary is an unordered collection of key-value pairs, where each key is unique. In Python, dictionaries are created using curly braces {} or the dict() function. Here's an example of creating a dictionary of name and age:

```
person = {'Alice': 25, 'Bob': 30, 'Charlie': 35}
```

In this example, person is a dictionary containing three key-value pairs, where the keys are the names 'Alice', 'Bob', and 'Charlie', and the values are their ages.

Accessing and Modifying Dictionary Elements

You can access and modify elements in a dictionary using the keys. To access the value associated with a key, you can use the square bracket [] notation. Here's an example:

```
person = {'Alice': 25, 'Bob': 30, 'Charlie': 35}
print(person['Bob']) # 30
```

In this example, we access the value associated with the key 'Bob' in the person dictionary.

To modify the value associated with a key, you can simply assign a new value using the square bracket [] notation.

 Here's an example:

```
person = {'Alice': 25, 'Bob': 30}
person.update({'Charlie': 35, 'David': 40})
print(person) # {'Alice': 25, 'Bob': 30, 'Charlie': 35, 'David': 40}
```

In this example, we add two new key-value pairs, 'Charlie': 35 and 'David': 40, to the person dictionary using the update() method.

To remove a key-value pair from a dictionary, you can use the del keyword or the pop() method. The del keyword removes the key-value pair associated with a given key, while the pop() method removes the key-value pair associated with a given key and returns its value. Here's an example:

```
person = {'Alice': 25, 'Bob': 30, 'Charlie': 35}
del person['Bob']
print(person) # {'Alice': 25, 'Charlie': 35}
```

```
age = person.pop('Alice')
print(age) # 25
print(person) # {'Charlie': 35}
```

In this example, we first remove the key-value pair associated with the key 'Bob' from the person dictionary using the del keyword. We then remove the key-value pair associated with the key 'Alice' from the person dictionary using the pop() method and assign its value to the variable age.

Sets and dictionaries are two important data structures in Python that can be used to store and manipulate collections of data. Sets are useful for storing unique elements and performing set operations such as union, intersection, and difference. Dictionaries are useful for storing key-value pairs and performing lookups based on keys. By understanding the basics of sets and dictionaries, you can write more efficient and expressive Python code.

Tuples and sequences

In Python, tuples and sequences are two important data structures that are used to store collections of elements. Tuples are similar to lists, but they are immutable, which means that their contents cannot be modified once they are created. Sequences are a more general concept that includes both tuples and lists. In this article, we will explore the basics of tuples and sequences in Python, including how to create them, access their elements, and perform operations on them.

Creating Tuples and Sequences

Tuples are created by enclosing a sequence of elements in parentheses, separated by commas. Here's an example:

```
t = (1, 2, 3)
print(t) # (1, 2, 3)
```

In this example, we create a tuple t containing the elements 1, 2, and 3.

Sequences can also be created using the tuple() function or by converting a list to a tuple using the tuple() function. Here are some examples:

```
t = tuple([1, 2, 3])
print(t) # (1, 2, 3)

t = tuple("hello")
print(t) # ('h', 'e', 'l', 'l', 'o')
```

In the first example, we create a tuple t containing the elements 1, 2, and 3 by converting a list to a tuple using the tuple() function. In the second example, we create a tuple t containing the characters of the string "hello".

Accessing Elements of Tuples and Sequences

Elements of tuples and sequences can be accessed using indexing and slicing. Indexing is used to access a single element of a tuple or sequence, while slicing is used to access a subsequence of elements. The syntax for indexing and slicing is the same as for lists. Here are some examples:

```
t = (1, 2, 3, 4, 5)
print(t[0]) # 1
print(t[-1]) # 5
```

```
print(t[1:3]) # (2, 3)
```

In this example, we create a tuple t containing the elements 1, 2, 3, 4, and 5. We then access the first element of t using the index 0, the last element of t using the index -1, and a subsequence of elements from the second element to the third element using slicing.

Performing Operations on Tuples and Sequences

Tuples and sequences can be used in the same way as lists to perform operations such as concatenation, repetition, and membership testing. However, because tuples are immutable, some operations such as sorting and appending are not available. Here are some examples:

```
t1 = (1, 2, 3)
t2 = (4, 5, 6)
t3 = t1 + t2
print(t3) # (1, 2, 3, 4, 5, 6)

t4 = t1 * 3
print(t4) # (1, 2, 3, 1, 2, 3, 1, 2, 3)

t5 = (1, 2, 3)
print(2 in t5) # True
print(4 in t5) # False
```

In this example, we create two tuples t1 and t2 containing the elements 1, 2, and 3 and 4, 5, and 6, respectively. We then concatenate t1 and `t2 using the + operator to create a new tuple t3 containing the elements 1, 2, 3, 4, 5, and 6. We also repeat t1 three times using the * operator to create a new tuple t4 containing the elements 1, 2, 3, 1, 2, 3, 1, and 2, 3. Finally, we test for membership in t5 using the in operator, which returns True if the element is in the tuple and False otherwise.

Tuples and sequences are useful data structures in Python for storing collections of elements. Tuples are immutable, which makes them useful for storing fixed data such as coordinates, while sequences are more general and can be used for storing any type of data. In this article, we covered the basics of creating tuples and sequences, accessing their elements, and performing operations on them.

- ## Data structures in Python

Data structures in Python are a fundamental concept for storing, organizing, and manipulating data. Python provides a wide range of built-in data structures, including lists, tuples, sets, dictionaries, and more. In this article, we will explore each of these data structures, their properties, and how they can be used in Python.

Lists

Lists are one of the most commonly used data structures in Python. A list is a collection of elements that can be of any type, including other lists. Lists are mutable, meaning that their elements can be added, removed, or modified. Here is an example of creating a list, accessing its elements, and modifying it:

```
# Creating a list
my_list = [1, 2, 3, 4, 5]

# Accessing elements of a list
print(my_list[0]) # 1
print(my_list[-1]) # 5

# Modifying elements of a list
my_list[0] = 0
my_list.append(6)
print(my_list) # [0, 2, 3, 4, 5, 6]
```

Tuples

Tuples are similar to lists but are immutable, meaning that their elements cannot be modified once they are created. Tuples are often used to store fixed data such as coordinates, dates, or phone numbers. Here is an example of creating a tuple, accessing its elements, and using tuple unpacking:

```
# Creating a tuple
my_tuple = (1, 2, 3)

# Accessing elements of a tuple
print(my_tuple[0]) # 1
print(my_tuple[-1]) # 3

# Tuple unpacking
```

```
x, y, z = my_tuple
print(x) # 1
print(y) # 2
print(z) # 3
```

Sets

Sets are unordered collections of unique elements. Sets can be used to perform set operations such as union, intersection, and difference. Here is an example of creating a set, adding and removing elements, and performing set operations:

```
# Creating a set
my_set = {1, 2, 3}

# Adding and removing elements of a set
my_set.add(4)
my_set.remove(1)
print(my_set) # {2, 3, 4}

# Set operations
my_set2 = {3, 4, 5}
print(my_set.union(my_set2)) # {2, 3, 4, 5}
print(my_set.intersection(my_set2)) # {3, 4}
print(my_set.difference(my_set2)) # {2}
```

Dictionaries

Dictionaries are collections of key-value pairs. Each key is associated with a value, which can be of any type. Dictionaries are often used to store data in a way that can be easily accessed using a unique identifier such as a name or an ID. Here is an example of creating a dictionary, accessing its values, and modifying it:

```
# Creating a dictionary
my_dict = {"name": "John", "age": 30}

# Accessing values of a dictionary
print(my_dict["name"]) # John
print(my_dict.get("age")) # 30

# Modifying values of a dictionary
my_dict["age"] = 31
my_dict["city"] = "New York"
print(my_dict) # {'name': 'John', 'age': 31, 'city': 'New York'}
```

Stack and Queue implementation in Python

Stacks and queues are two important abstract data types commonly used in computer science. They are used to store and manipulate data in various ways. In this article, we will explore the implementation of stacks and queues in Python.

Stacks

A stack is a collection of elements that supports two main operations: push and pop. Push adds an element to the top of the stack, and pop removes the top element from the stack. The last element added to the stack is the first element to be removed (LIFO: Last In, First Out). Stacks are used in various applications such as expression evaluation, backtracking, and parsing.

In Python, we can implement a stack using a list. Here is an example of a stack implementation in Python:

```python
class Stack:
    def __init__(self):
        self.stack = []

    def push(self, item):
        self.stack.append(item)

    def pop(self):
        return self.stack.pop()

    def is_empty(self):
        return len(self.stack) == 0

    def peek(self):
        return self.stack[-1]
```

In the above code, we have defined a class called Stack that has four methods: push, pop, is_empty, and peek. The push method adds an element to the top of the stack, the pop method removes the top element from the stack, the is_empty method checks whether the stack is empty, and the peek method returns the top element of the stack without removing it.

Queues

A queue is a collection of elements that supports two main operations: enqueue and dequeue. Enqueue adds an element to the rear of the queue, and dequeue removes an element from the front of the queue. The first element added to the queue is the first element to be removed

(FIFO: First In, First Out). Queues are used in various applications such as scheduling, resource allocation, and event handling.

In Python, we can implement a queue using a list. However, this is not very efficient because adding or removing elements from the front of the list requires shifting all the other elements. Therefore, it is better to use a deque (double-ended queue) from the collections module. Here is an example of a queue implementation using a deque:

```python
from collections import deque

class Queue:
    def __init__(self):
        self.queue = deque()

    def enqueue(self, item):
        self.queue.append(item)

    def dequeue(self):
        return self.queue.popleft()

    def is_empty(self):
        return len(self.queue) == 0

    def peek(self):
        return self.queue[0]
```

In the above code, we have defined a class called Queue that has four methods: enqueue, dequeue, is_empty, and peek. The enqueue method adds an element to the rear of the queue, the dequeue method removes an element from the front of the queue, the is_empty method checks whether the queue is empty, and the peek method returns the front element of the queue without removing it.

In this article, we have explored the implementation of stacks and queues in Python. Stacks are useful for storing and manipulating data in a Last In, First Out (LIFO) manner, while queues are useful for storing and manipulating data in a First In, First Out (FIFO) manner. By using the code examples provided, you can easily implement these data structures in your own Python projects.

Binary trees and graphs in Python

Binary trees and graphs are important data structures used in computer science and are used to represent and manipulate various types of data. In this article, we will explore the implementation of binary trees and graphs in Python.

Binary Trees

A binary tree is a tree data structure where each node has at most two child nodes, referred to as the left and right child nodes. The first node of the tree is known as the root node. Binary trees are used in various applications such as searching and sorting algorithms, expression evaluation, and decision making.

In Python, we can implement a binary tree using classes. Here is an example of a binary tree implementation in Python:

```python
class Node:
    def __init__(self, value):
        self.value = value
        self.left_child = None
        self.right_child = None

class BinaryTree:
    def __init__(self, root):
        self.root = Node(root)

    def print_tree(self, traversal_type):
        if traversal_type == "inorder":
            return self.inorder_traversal(self.root, "")
        elif traversal_type == "preorder":
            return self.preorder_traversal(self.root, "")
        elif traversal_type == "postorder":
            return self.postorder_traversal(self.root, "")
        else:
            print("Traversal type " + str(traversal_type) + " is not supported.")

    def inorder_traversal(self, start, traversal):
        if start:
            traversal = self.inorder_traversal(start.left_child, traversal)
            traversal += (str(start.value) + "-")
            traversal = self.inorder_traversal(start.right_child, traversal)
        return traversal

    def preorder_traversal(self, start, traversal):
        if start:
```

```
        traversal += (str(start.value) + "-")
        traversal = self.preorder_traversal(start.left_child, traversal)
        traversal = self.preorder_traversal(start.right_child, traversal)
    return traversal

def postorder_traversal(self, start, traversal):
    if start:
        traversal = self.postorder_traversal(start.left_child, traversal)
        traversal = self.postorder_traversal(start.right_child, traversal)
        traversal += (str(start.value) + "-")
    return traversal
```

In the above code, we have defined two classes, Node and BinaryTree. The Node class represents a single node in the binary tree and has three attributes: value, left_child, and right_child. The BinaryTree class represents the binary tree and has four methods: print_tree, inorder_traversal, preorder_traversal, and postorder_traversal. The print_tree method is used to print the tree in a specific order, the inorder_traversal method traverses the tree in order, the preorder_traversal method traverses the tree in pre-order, and the postorder_traversal method traverses the tree in post-order.

Graphs

A graph is a collection of vertices (nodes) and edges that connect these vertices. Graphs are used in various applications such as social networks, transportation networks, and computer networks.

In Python, we can implement a graph using a dictionary. Here is an example of a graph implementation in Python:

```
class Graph:
    def __init__(self):
        self.graph_dict = {}

    def add_vertex(self, vertex):
        if vertex not in self.graph_dict:
            self.graph_dict[vertex] = []

    def add_edge(self, vertex1, vertex2):
        if vertex1 in self.graph_dict:
            self.graph_dict[vertex1].append(vertex2)
        else:
            self.graph_dict[vertex1] = [vertex2]

    def display_graph(self):
        print(self.graph_dict
```

In the above code, we have defined a class Graph that has three methods: add_vertex, add_edge, and display_graph. The add_vertex method is used to add a vertex to the graph, the add_edge method is used to add an edge between two vertices, and the display_graph method is used to display the graph.

Here is an example of how to use the Graph class to create and display a graph:

```
g = Graph()
g.add_vertex('A')
g.add_vertex('B')
g.add_vertex('C')
g.add_vertex('D')
g.add_vertex('E')

g.add_edge('A', 'B')
g.add_edge('B', 'C')
g.add_edge('C', 'D')
g.add_edge('D', 'E')
g.add_edge('E', 'A')

g.display_graph()
```

Output:

{'A': ['B', 'E'], 'B': ['C'], 'C': ['D'], 'D': ['E'], 'E': ['A']}

In the above example, we first create an instance of the Graph class and then add five vertices to it using the add_vertex method. We then add five edges between these vertices using the add_edge method. Finally, we display the graph using the display_graph method.

In this article, we explored the implementation of binary trees and graphs in Python. Binary trees are used to represent and manipulate various types of data, and graphs are used to represent the relationships between different objects. With the help of the code examples provided, you can easily implement binary trees and graphs in your Python programs.

Recursion in Python

Recursion in Python: A Comprehensive Guide to Writing Recursive Functions

Recursion is a powerful concept in computer programming that allows a function to call itself. In Python, recursion is often used to solve problems that can be broken down into smaller, simpler problems. In this article, we will explore the basics of recursion, how it works in Python, and provide examples of recursive functions.

What is Recursion?

Recursion is a technique where a function calls itself to solve a problem. When a function calls itself, it creates a new instance of that function with its own set of variables. The new instance of the function then runs, and can call itself again if necessary. This process continues until the function reaches a base case, where it no longer needs to call itself.

A recursive function consists of two parts: the base case and the recursive case. The base case is the terminating condition that stops the recursion. The recursive case is the condition that calls the function again with a simpler version of the problem.

Recursive Functions in Python

In Python, a recursive function is defined like any other function, with the addition of a recursive call to itself. For example, let's say we want to write a function that calculates the factorial of a number using recursion. The factorial of a number is the product of all positive integers up to and including that number.

We can define the factorial function recursively as follows:

```
def factorial(n):
    if n == 1:
        return 1
    else:
        return n * factorial(n-1)
```

In this function, the base case is when n is equal to 1. When n equals 1, the function returns 1, which stops the recursion. If n is not equal to 1, the function calls itself with n-1 as the argument. The result of this call is multiplied by n and returned.

To see how this function works, let's call it with the argument 5:

```
factorial(5)
```

The function first checks if 5 is equal to 1. Since it is not, the function calls itself with the argument 4. The new instance of the function checks if 4 is equal to 1, and since it is not, it calls itself with the argument 3. This process continues until the function reaches the base case, where n equals 1. The function then returns the result of the last multiplication, which is 5 * 4 * 3 * 2 * 1, or 120.

Another example of a recursive function is the Fibonacci sequence. The Fibonacci sequence is a series of numbers in which each number is the sum of the two preceding numbers. The first two numbers in the sequence are 0 and 1.

We can define the Fibonacci sequence recursively as follows:

```python
def fibonacci(n):
    if n == 0:
        return 0
    elif n == 1:
        return 1
    else:
        return fibonacci(n-1) + fibonacci(n-2)
```

In this function, the base cases are when n is equal to 0 or 1. If n is equal to 0, the function returns 0. If n is equal to 1, the function returns 1. If n is not equal to 0 or 1, the function calls itself twice with n-1 and n-2 as the arguments. The results of these calls are added together and returned.

To see how this function works, let's call it with the argument 6:

fibonacci(6)

The function first checks if 6 is equal to 0 or 1. Since it is not, the function calls itself twice with the arguments 5 and 4. The new instances of the function call themselves with smaller values until they reach the base cases of 0 or 1. The results of these calls are added together and returned up the chain until the original call to fibonacci(6) returns the value 8.

Common Pitfalls with Recursion

Recursion can be a powerful tool in Python, but it can also lead to some common pitfalls. One of the most common pitfalls is the risk of infinite recursion. This occurs when a function never reaches its base case and continues to call itself indefinitely. This can cause the program to crash or run out of memory.

To avoid infinite recursion, it is important to ensure that each recursive call moves the function closer to the base case. In addition, it is important to choose appropriate base cases to ensure that the function will eventually terminate.

Another pitfall is the risk of stack overflow. Since each recursive call creates a new instance of the function, the function call stack can quickly become very deep. This can cause the program to run out of stack memory and crash.

To avoid stack overflow, it is important to use tail recursion. Tail recursion occurs when the final instruction of a function is a recursive call. In this case, the compiler or interpreter can optimize the code to avoid creating new stack frames, which reduces the risk of stack overflow.

Recursion is a powerful tool in Python that allows functions to call themselves to solve problems. Recursive functions consist of a base case and a recursive case, and are often used to solve problems that can be broken down into smaller, simpler problems.

However, recursion can also lead to common pitfalls such as infinite recursion and stack overflow. To avoid these pitfalls, it is important to ensure that each recursive call moves the function closer to the base case, and to use tail recursion when possible.

By understanding the basics of recursion and how to avoid common pitfalls, you can use this powerful technique to solve complex problems in your Python programs.

Debugging in Python

Debugging in Python: Tips and Techniques for Finding and Fixing Bugs

Debugging is an essential part of software development. Even the most experienced programmers make mistakes, and it is important to be able to find and fix bugs in your code. In this article, we will explore some tips and techniques for debugging in Python.

Understanding Errors in Python

Before we dive into the tips and techniques for debugging, it is important to understand the types of errors that can occur in Python.

Syntax errors occur when the Python interpreter cannot parse your code. These errors are often caused by typos, missing parentheses, or other syntax errors. Python will raise a SyntaxError and provide a message indicating where the error occurred.

```
# Example of a syntax error
print "Hello World!"
```

Name errors occur when you try to use a variable or function that has not been defined. Python will raise a NameError and provide a message indicating which name is not defined.

```
# Example of a name error
print(x)
```

Type errors occur when you try to use an object of the wrong type. Python will raise a TypeError and provide a message indicating which types are involved.

```
# Example of a type error
print("The answer is " + 42)
```

Debugging Techniques in Python

Now that we understand the types of errors that can occur in Python, let's explore some tips and techniques for debugging.

Use print statements

One of the simplest and most effective debugging techniques is to use print statements to output the value of variables at various points in your code. This can help you understand how your code is executing and identify where the problem is occurring.

```
# Example of using print statements
x = 10
y = 20
print("x = ", x)
print("y = ", y)
result = x + y
print("result = ", result)
```

Use a debugger

Python provides a built-in debugger called pdb that can be used to step through your code and identify where the problem is occurring. The debugger allows you to set breakpoints, view the value of variables, and execute code line by line.

To use the debugger, you can import the pdb module and call the set_trace() function at the point where you want to start debugging.

```
# Example of using the pdb debugger
import pdb

def multiply(x, y):
    result = x * y
    pdb.set_trace()
    return result

multiply(2, 3)
```

When the set_trace() function is called, the debugger will pause execution of the code and provide a prompt where you can enter commands to inspect variables and step through the code.

Use assertions

Assertions are statements that check whether a condition is true and raise an AssertionError if the condition is false. Assertions can be used to check that the input to a function is valid or that a particular condition holds true at a certain point in your code.

```
# Example of using assertions
def divide(x, y):
    assert y != 0, "Cannot divide by zero"
    return x / y

result = divide(10, 2)
print(result)
result = divide(10, 0)
```

In this example, the assert statement checks whether y is not equal to 0 and raises an AssertionError if it is. This can help you identify errors in your code and provide helpful error messages for users.

Debugging is an essential part of software development, and Python provides a variety of tools and techniques for finding and fixing bugs in your code. By using print statements, debuggers, and assertions, you can identify and resolve errors in your code and improve the quality of your software.

Testing in Python

Testing in Python: Techniques for Ensuring Code Quality

As a software developer, it is important to ensure that your code is high-quality and free of bugs. One of the best ways to do this is through testing. In this article, we will explore the basics of testing in Python, including techniques for writing and executing tests.

Why Test Your Code?

Testing is a critical part of the software development process. It helps ensure that your code is functioning correctly and that it will continue to function as expected even as you make changes to it. Without testing, bugs and errors can go undetected, leading to more serious problems down the line.

Types of Tests in Python

There are several types of tests that can be performed in Python:

Unit Tests: These tests focus on individual units or functions in your code. They are designed to test the behavior of a single piece of code in isolation.

Integration Tests: These tests focus on the interaction between different components or modules in your code. They are designed to test how these components work together.

Functional Tests: These tests focus on the behavior of your code from the perspective of the user. They are designed to test the entire system, from input to output.

Writing Tests in Python

To write tests in Python, you will typically use a testing framework such as pytest or unittest. These frameworks provide a set of tools and functions for writing and executing tests.

Here is an example of a simple test using pytest:

```
# Example of a test using pytest
def test_addition():
    assert 2 + 2 == 4
```

In this example, we define a test called test_addition that checks whether 2 + 2 equals 4. We use the assert statement to check this condition and pytest will report whether the test passed or failed.

Executing Tests in Python

Once you have written your tests, you can execute them using the testing framework. To do this, you typically run a command such as pytest or python -m unittest.

Here is an example of running tests using pytest:

In this example, we run the pytest command and specify the name of the file containing our tests.

Best Practices for Testing in Python

To ensure that your tests are effective and useful, there are several best practices to keep in mind:

Write tests for all of your code: Make sure that you write tests for every function or module in your code.

Use descriptive test names: Use descriptive names for your tests so that it is clear what they are testing.

Test edge cases: Make sure that your tests cover edge cases and unusual input values.

Run tests frequently: Make sure that you run your tests frequently, especially after making changes to your code.

Testing is an essential part of the software development process, and Python provides a variety of tools and frameworks for writing and executing tests. By following best practices and writing comprehensive tests, you can ensure that your code is high-quality and free of bugs.

Debugging and profiling Python applications

Debugging and Profiling Python Applications: Tips and Techniques

Debugging and profiling are essential tools for any software developer. In this article, we will explore some tips and techniques for debugging and profiling Python applications.

Debugging Python Applications

Debugging is the process of identifying and fixing errors in your code. Python provides several tools and techniques for debugging your applications.

Using print statements: One of the simplest ways to debug your code is by using print statements. By adding print statements to your code, you can see the values of variables and the flow of your program.

```
# Example of using print statements for debugging
def add(a, b):
    print("Adding", a, "and", b)
    result = a + b
    print("Result:", result)
    return result
```

Using pdb: The Python Debugger (pdb) is a powerful tool for debugging your code. It allows you to step through your code line by line and examine the values of variables.

```
# Example of using pdb for debugging
import pdb

def add(a, b):
    pdb.set_trace()
    result = a + b
    return result
```

Using logging: The logging module in Python allows you to log messages at different levels of severity. By using logging, you can easily trace the flow of your program and identify errors.

```
# Example of using logging for debugging
import logging

def add(a, b):
    logging.debug("Adding %s and %s", a, b)
    result = a + b
    logging.debug("Result: %s", result)
    return result
```

Profiling Python Applications

Profiling is the process of analyzing the performance of your code. It allows you to identify the parts of your code that are slow and optimize them.

Using cProfile: The cProfile module in Python provides a way to profile your code at the function level. It allows you to see how much time is spent in each function and how many times each function is called.

```
# Example of using cProfile for profiling
import cProfile

def main():
    # Code to be profiled goes here
    pass

if __name__ == '__main__':
    cProfile.run('main()')
```

Using line_profiler: The line_profiler module in Python provides a way to profile your code at the line level. It allows you to see how much time is spent on each line of code.

```
# Example of using line_profiler for profiling
!pip install line_profiler

%load_ext line_profiler

@profile
def add(a, b):
    result = a + b
    return result
```

Using memory_profiler: The memory_profiler module in Python provides a way to profile the memory usage of your code. It allows you to see how much memory is used by each line of code.

```
# Example of using memory_profiler for profiling
!pip install memory_profiler

%load_ext memory_profiler

@profile
def add(a, b):
    result = a + b
    return result
```

Debugging and profiling are essential tools for any software developer. Python provides several tools and techniques for debugging and profiling your applications. By using these tools and techniques, you can identify and fix errors in your code and optimize its performance.

Virtual Environments in Python: A Comprehensive Guide

Python is a popular programming language used in various fields such as web development, data science, and machine learning. When developing applications using Python, it is essential to maintain the dependencies and versions of the packages used. Virtual environments provide a solution to this problem by creating an isolated environment for each project. This guide will cover everything you need to know about virtual environments in Python.

What are Virtual Environments?

Virtual environments are isolated environments created specifically for a project. They allow developers to install specific versions of Python and packages without affecting the system-level Python installation. This isolation prevents conflicts between packages and enables the creation of reproducible development environments.

Creating Virtual Environments

Python comes with a built-in module for creating virtual environments called venv. Here's how you can create a virtual environment:

```
python -m venv myenv
```

The above command creates a virtual environment named myenv. Once created, you can activate the virtual environment using the following command:

```
source myenv/bin/activate
```

Managing Packages in Virtual Environments

Once you have activated the virtual environment, you can install packages just like you would in a system-level Python installation. However, the packages will be installed only in the virtual environment and not in the system-level installation. Here's an example of installing the NumPy package in a virtual environment:

```
pip install numpy
```

Freezing Requirements

It's a good practice to freeze the requirements of your project, including the version of Python and packages used, to ensure reproducibility. You can do this using the pip freeze command:

pip freeze > requirements.txt

This command creates a file named requirements.txt that contains a list of all the packages installed in the virtual environment along with their versions.

Sharing Virtual Environments

Virtual environments can be shared with other developers to ensure that they can reproduce your development environment. You can share your virtual environment by creating a requirements.txt file and sharing it along with the codebase.

Here's how another developer can create the same virtual environment as yours:

```
python -m venv myenv
source myenv/bin/activate
pip install -r requirements.txt
```

Managing Multiple Virtual Environments

When working on multiple projects, it's essential to maintain separate virtual environments for each project. You can create multiple virtual environments by specifying a different name for each environment:

```
python -m venv project1
python -m venv project2
```

To activate a specific virtual environment, use the source command with the path to the virtual environment's activation script:

```
source project1/bin/activate
```

Virtual environments provide an efficient and practical way to manage dependencies and versions of packages used in a Python project. They enable the creation of isolated development environments that are reproducible and free from package conflicts. With the venv module, creating and managing virtual environments is easy and straightforward. By using virtual environments, you can ensure that your Python applications are portable and can be easily shared with other developers.

Concurrency in Python: A Practical Guide

Concurrency is the ability of a program to execute multiple tasks simultaneously. Python provides several mechanisms for achieving concurrency, including threading, multiprocessing, and asynchronous programming. In this guide, we will explore each of these mechanisms and discuss when to use each one.

Threading

Threading is a mechanism for achieving concurrency by running multiple threads within a single process. Each thread can execute a different task simultaneously, allowing for improved performance and responsiveness. Here's an example of using threading in Python:

```python
import threading

def worker():
    """Thread worker function"""
    print('Worker thread started')
    # Do some work here
    print('Worker thread finished')

# Create a new thread
thread = threading.Thread(target=worker)
# Start the thread
thread.start()

# Wait for the thread to finish
thread.join()

print('Main thread finished')
```

The above code creates a new thread and starts it using the start() method. The join() method is called to wait for the thread to finish before continuing execution.

Multiprocessing

Multiprocessing is a mechanism for achieving concurrency by running multiple processes in parallel. Each process runs in its own memory space, providing isolation and security. Here's an example of using multiprocessing in Python:

```python
import multiprocessing

def worker():
    """Process worker function"""
    print('Worker process started')
    # Do some work here
    print('Worker process finished')

# Create a new process
process = multiprocessing.Process(target=worker)
# Start the process
process.start()

# Wait for the process to finish
process.join()

print('Main process finished')
```

The above code creates a new process and starts it using the start() method. The join() method is called to wait for the process to finish before continuing execution.

Asynchronous Programming

Asynchronous programming is a mechanism for achieving concurrency by allowing a program to perform multiple tasks simultaneously without creating additional threads or processes. This is achieved using coroutines and event loops. Here's an example of using asynchronous programming in Python:

```python
import asyncio

async def worker():
    """Coroutine worker function"""
    print('Worker coroutine started')
    # Do some work here
    print('Worker coroutine finished')

# Create a new event loop
loop = asyncio.get_event_loop()

# Schedule the coroutine to run
loop.run_until_complete(worker())

print('Main coroutine finished')
```

The above code creates a new event loop and schedules a coroutine to run using the run_until_complete() method. The event loop manages the execution of the coroutine and switches between tasks as needed.

Concurrency is an essential feature of modern programming languages and is essential for building high-performance applications. Python provides several mechanisms for achieving concurrency, including threading, multiprocessing, and asynchronous programming. Each mechanism has its own advantages and disadvantages, and the choice of mechanism will depend on the specific requirements of your application. By understanding the different mechanisms for achieving concurrency in Python, you can build more efficient and responsive applications.

Multiprocessing in Python: A Comprehensive Guide

Multiprocessing is a technique for parallel processing in which multiple processes are spawned to perform a task. In Python, the multiprocessing module provides a way to create and manage child processes that can run concurrently with the main process. This can be a powerful tool for speeding up computation-heavy tasks and achieving true parallelism.

In this guide, we will explore the multiprocessing module in Python and demonstrate how it can be used to improve performance in your Python applications.

Chapter 1: Getting Started with Multiprocessing

The multiprocessing module provides a way to create and manage child processes in Python. To get started, we first need to import the module:

```
import multiprocessing
```

Next, we can create a new process using the Process class:

```
def my_function():
    print('Hello from child process!')

process = multiprocessing.Process(target=my_function)
process.start()
```

In the above code, we define a function my_function() which will be run by the child process. We then create a new process using the Process class and pass our function as the target. Finally, we start the process using the start() method.

Chapter 2: Communication Between Processes

When using multiprocessing, it's often necessary to share data between the main process and child processes. The multiprocessing module provides several ways to achieve this, including pipes, queues, and shared memory.

Here's an example of using a Queue to share data between a parent and child process:

```
def my_function(queue):
    data = queue.get()
    print('Received:', data)

queue = multiprocessing.Queue()
queue.put('Hello from parent process!')
```

```
process = multiprocessing.Process(target=my_function, args=(queue,))
process.start()
```

In the above code, we create a Queue object and put some data into it. We then pass the queue as an argument to our child process function, my_function(). The child process retrieves the data from the queue using the get() method.

Chapter 3: Parallel Processing with Pool

The multiprocessing module provides a Pool class which can be used to create a pool of worker processes that can execute tasks in parallel. Here's an example of using Pool to execute a function on a list of inputs:

```
def my_function(input):
    return input ** 2

inputs = [1, 2, 3, 4, 5]
pool = multiprocessing.Pool()
results = pool.map(my_function, inputs)
```

In the above code, we define a function my_function() which takes an input and returns its square. We then create a list of inputs and a new Pool object. Finally, we use the map() method of the Pool object to execute our function on each input in parallel.

Chapter 4: Synchronization with Locks and Semaphores

When using multiprocessing, it's important to ensure that multiple processes don't access shared resources simultaneously. The multiprocessing module provides two synchronization primitives, locks and semaphores, which can be used to achieve this.

Here's an example of using a lock to synchronize access to a shared variable:

```
def my_function(lock, shared_variable):
    lock.acquire()
    shared_variable.value += 1
    lock.release()

lock = multiprocessing.Lock()
shared_variable = multiprocessing.Value('i', 0)

processes = []
for i in range(10):
    process = multiprocessing.Process(target=my_function, args=(lock, shared_variable))
    processes.append(process)
    process.start()
```

```
for process in processes:
    process.join()

print('Final value:', shared_variable.value)
```

In the above code, we define a function my_function() which increments a shared variable.

We create a new lock object using the Lock class and a shared variable using the Value class. We then create ten processes, each of which executes my_function() with the lock and shared variable as arguments. Finally, we wait for all processes to complete and print the final value of the shared variable.

Chapter 5: Process Pools with the multiprocessing module

The multiprocessing module also provides a Pool class that can be used to create a pool of worker processes that can be used to execute tasks in parallel.

```
from multiprocessing import Pool

def my_function(x):
    return x**2

if __name__ == '__main__':
    with Pool(processes=4) as pool:
        results = pool.map(my_function, range(10))
    print(results)
```

In the above code, we create a new Pool object with four worker processes using the with statement. We then use the map() method to apply the my_function() function to the range of integers from 0 to 9, in parallel. Finally, we print the results.

Chapter 6: Using the concurrent.futures module

The concurrent.futures module is another option for creating parallel processes in Python. This module provides a high-level interface for asynchronously executing functions using threads or processes.

```
from concurrent.futures import ProcessPoolExecutor

def my_function(x):
    return x**2

if __name__ == '__main__':
    with ProcessPoolExecutor(max_workers=4) as executor:
        results = executor.map(my_function, range(10))
```

```
print(list(results))
```

In the above code, we use the ProcessPoolExecutor class to create a pool of four worker processes. We then use the map() method to apply the my_function() function to the range of integers from 0 to 9, in parallel. Finally, we print the results.

Chapter 7: Conclusion

In this guide, we have explored the multiprocessing module in Python and demonstrated how it can be used to improve performance in your Python applications. We have covered a range of topics, including communication between processes, parallel processing with Pool, and synchronization with locks and semaphores.

By leveraging multiprocessing, you can make your Python applications more efficient and take full advantage of modern hardware. Whether you're working with large datasets or building complex simulations, multiprocessing is a powerful tool that can help you get the job done faster.

Threading in Python

Chapter 1: Introduction to Threading in Python

Python provides several ways to implement concurrency and parallelism in your programs. One of the most popular ways is through threading. Threading allows you to execute multiple threads of execution within a single process.

In this guide, we will explore threading in Python and learn how to use it to make your applications more efficient.

Chapter 2: The threading module

The threading module is the primary module for threading in Python. It provides a simple and efficient way to create and manage threads.

```
import threading

def my_function():
    print('Hello from thread', threading.current_thread().name)

if __name__ == '__main__':
    thread = threading.Thread(target=my_function)
    thread.start()
    print('Hello from main thread')
    thread.join()
```

In the above code, we create a new thread using the Thread class and the target parameter. We then start the thread using the start() method and wait for it to complete using the join() method. Finally, we print a message from the main thread and the thread we created.

Chapter 3: Sharing Data between Threads

Threads can share data between each other by using shared variables. However, if multiple threads try to access and modify the same variable simultaneously, it can lead to race conditions and data corruption.

The threading module provides a few synchronization primitives, including locks and semaphores, that can be used to prevent race conditions and ensure data consistency.

```
import threading

def increment(counter, lock):
    for _ in range(100000):
        with lock:
            counter.value += 1

if __name__ == '__main__':
    from multiprocessing import Value, Lock
    counter = Value('i', 0)
    lock = Lock()
    threads = [threading.Thread(target=increment, args=(counter, lock)) for _ in range(4)]
    for thread in threads:
        thread.start()
    for thread in threads:
        thread.join()
    print('Counter:', counter.value)
```

In the above code, we create a new Value object and a Lock object from the multiprocessing module. We then create four threads and pass the shared Value and Lock objects as arguments to the increment() function. We use the with statement to acquire and release the lock around the critical section of the code that modifies the shared variable.

Chapter 4: Threading with Queues

Queues can be used to coordinate work between threads. The queue module provides the Queue class, which is a thread-safe implementation of a queue.

```
import queue
import threading

def worker(queue):
    while True:
        item = queue.get()
        if item is None:
            break
        print(item)

if __name__ == '__main__':
    q = queue.Queue()
    threads = [threading.Thread(target=worker, args=(q,)) for _ in range(4)]
    for thread in threads:
        thread.start()
    for i in range(10):
        q.put(i)
    for _ in range(4):
        q.put(None)
```

```
for thread in threads:
    thread.join()
```

In the above code, we create a new Queue object and four worker threads that consume items from the queue. We use the put() method to add items to the queue and the None object to signal the end of the work. We then use the join() method to wait for all threads to complete.

Asynchronous programming with asyncio

Chapter 1: Introduction to Asynchronous Programming with asyncio

Asynchronous programming is a way to write programs that can perform multiple tasks concurrently, without blocking the execution of the program. Python provides several ways to implement asynchronous programming, including asyncio.

In this guide, we will explore asyncio in Python and learn how to use it to make your applications more efficient.

Chapter 2: The asyncio module

The asyncio module is the primary module for asynchronous programming in Python. It provides a simple and efficient way to create and manage asynchronous tasks.

```
import asyncio

async def my_coroutine():
    print('Hello from coroutine')

if __name__ == '__main__':
    loop = asyncio.get_event_loop()
    loop.run_until_complete(my_coroutine())
    loop.close()
```

In the above code, we create a new coroutine using the async keyword and the async def syntax. We then use the run_until_complete() method to execute the coroutine and the close() method to close the event loop.

Chapter 3: Asynchronous I/O

Asynchronous I/O is a way to perform I/O operations without blocking the execution of the program. The asyncio module provides several functions for performing I/O operations asynchronously, including asyncio.open_connection() and asyncio.start_server().

```
import asyncio

async def handle_client(reader, writer):
    while True:
        data = await reader.read(1024)
        if not data:
```

```
            break
        writer.write(data)
        await writer.drain()
    writer.close()

if __name__ == '__main__':
    loop = asyncio.get_event_loop()
    server = loop.run_until_complete(asyncio.start_server(handle_client, 'localhost', 8888))
    try:
        loop.run_forever()
    except KeyboardInterrupt:
        pass
    server.close()
    loop.run_until_complete(server.wait_closed())
    loop.close()
```

In the above code, we create a server using the start_server() function and the handle_client() coroutine. We use the await keyword to perform asynchronous I/O operations on the client connection, including reading data and writing data back to the client.

Chapter 4: Asynchronous Tasks

Asynchronous tasks are a way to perform multiple tasks concurrently in an asynchronous program. The asyncio module provides several functions for creating and managing asynchronous tasks, including asyncio.create_task() and asyncio.gather().

```
import asyncio

async def my_task(name):
    print(f'Task {name} started')
    await asyncio.sleep(1)
    print(f'Task {name} finished')

if __name__ == '__main__':
    loop = asyncio.get_event_loop()
    tasks = [loop.create_task(my_task(i)) for i in range(5)]
    loop.run_until_complete(asyncio.gather(*tasks))
    loop.close()
```

In the above code, we create five asynchronous tasks using the create_task() function and the my_task() coroutine. We use the gather() function to wait for all tasks to complete before closing the event loop.

In this guide, we have explored asyncio in Python and learned how to use the asyncio module to create and manage asynchronous tasks and perform asynchronous I/O operations.

Asynchronous programming can make your applications more efficient and responsive, and asyncio provides a simple and efficient way to implement it in Python.

Networking in Python

Chapter 1: Introduction to Networking in Python

Python provides powerful networking capabilities that allow you to build a wide range of network applications. With Python, you can create applications that communicate over the internet, access remote resources, and share data between machines.

In this guide, we will explore networking in Python and learn how to use it to build robust and scalable network applications.

Chapter 2: Socket Programming in Python

Socket programming is a low-level networking interface that provides a way to communicate between computers over a network. Python provides a rich set of libraries for socket programming, including the socket module.

```python
import socket

HOST = 'localhost'
PORT = 5000

with socket.socket(socket.AF_INET, socket.SOCK_STREAM) as s:
    s.bind((HOST, PORT))
    s.listen()
    conn, addr = s.accept()
    with conn:
        print(f'Connected by {addr}')
        while True:
            data = conn.recv(1024)
            if not data:
                break
            conn.sendall(data)
```

In the above code, we create a socket object using the socket.socket() function and the AF_INET and SOCK_STREAM constants. We then bind the socket to a host and port, listen for incoming connections, and accept the connection. We use the recv() and sendall() methods to receive and send data over the socket.

Chapter 3: HTTP Programming in Python

HTTP is the protocol used for communicating between web clients and servers over the internet. Python provides several libraries for working with HTTP, including the http.client and urllib modules.

```python
import http.client

conn = http.client.HTTPSConnection('www.python.org')
conn.request('GET', '/')
response = conn.getresponse()
print(response.status, response.reason)
data = response.read()
conn.close()
```

In the above code, we create an HTTPS connection to the Python website using the http.client.HTTPSConnection() function. We then send a GET request to retrieve the homepage and read the response data.

Chapter 4: Asynchronous Networking with asyncio

Asynchronous networking allows you to perform multiple network operations concurrently in a single thread, making your applications more efficient and responsive. Python provides several libraries for asynchronous networking, including the asyncio module.

```python
import asyncio
import aiohttp

async def fetch(session, url):
    async with session.get(url) as response:
        return await response.text()

async def main():
    async with aiohttp.ClientSession() as session:
        html = await fetch(session, 'http://www.python.org')
        print(html)

if __name__ == '__main__':
    loop = asyncio.get_event_loop()
    loop.run_until_complete(main())
```

In the above code, we create an asynchronous HTTP client using the aiohttp library and the ClientSession object. We use the async with syntax to manage the session and the fetch() coroutine to perform an HTTP GET request and return the response text. We use the run_until_complete() method to run the main() coroutine.

Chapter 5: Conclusion

In this guide, we have explored networking in Python and learned how to use socket programming, HTTP programming, and asynchronous networking to build robust and scalable network applications. Python provides a rich set of libraries for networking, making it easy to create applications that communicate over the internet and share data between machines.

Working with APIs in Python

Introduction to APIs

Application Programming Interfaces (APIs) provide a way for software applications to communicate with each other. APIs allow developers to access and manipulate data from other software applications or services, making it easy to integrate different systems and build more powerful applications.

In this guide, we will explore how to work with APIs in Python and learn how to use them to build robust and scalable applications.

RESTful APIs

RESTful APIs are a popular type of API that use HTTP methods to perform operations on resources. REST APIs are stateless and use URIs to represent resources, making them easy to use and understand. Python provides several libraries for working with RESTful APIs, including the requests and http.client modules.

```
import requests

response = requests.get('https://api.github.com/users/username/repos')
print(response.json())
```

In the above code, we use the requests library to send a GET request to the GitHub API and retrieve information about the repositories owned by a user. We use the json() method to convert the response data to a Python object.

Authentication and Authorization

Many APIs require authentication and authorization to access their resources. Python provides several libraries for working with authentication, including the requests and oauthlib modules.

```
import requests
from requests_oauthlib import OAuth1

url = 'https://api.twitter.com/1.1/account/verify_credentials.json'
auth = OAuth1('API_KEY', 'API_SECRET', 'ACCESS_TOKEN', 'ACCESS_TOKEN_SECRET')
response = requests.get(url, auth=auth)
print(response.json())
```

In the above code, we use the requests_oauthlib library to authenticate with the Twitter API and retrieve information about the authenticated user. We use the OAuth1 class to provide the API key, secret, access token, and access token secret.

Using APIs with Data Science Libraries

Python provides powerful libraries for data science and machine learning, such as numpy, pandas, and scikit-learn. These libraries can be used with APIs to retrieve and analyze data.

```
import requests
import pandas as pd

response = requests.get('https://api.coinmarketcap.com/v1/ticker/')
data = response.json()
df = pd.DataFrame(data)
print(df.head())
```

In the above code, we use the requests library to send a GET request to the CoinMarketCap API and retrieve information about cryptocurrency prices. We use the json() method to convert the response data to a Python object, and then create a pandas DataFrame to analyze the data.

In this guide, we have explored how to work with APIs in Python and learned how to use RESTful APIs, authentication and authorization, and data science libraries to build robust and scalable applications. APIs are a powerful tool for integrating different systems and building more powerful applications, and Python provides a rich set of libraries for working with APIs.

Web development in Python

Python is a versatile programming language that has been used for a wide range of applications, including web development. With its clean syntax, extensive libraries, and vast community, Python has become a popular choice for building web applications. In this news article, we will explore the latest developments in web development using Python.

Introduction to Web Development in Python

Web development involves creating websites, web applications, and web services that are accessed over the internet. Python provides several libraries and frameworks for web development, including Flask, Django, and Pyramid. These libraries and frameworks make it easy to build web applications quickly and efficiently.

Flask

Flask is a lightweight and flexible web framework for Python. Flask provides a simple and elegant way to create web applications and APIs. Flask comes with a built-in web server, making it easy to get started. Flask also supports a variety of extensions that can add functionality to your application, such as authentication, database integration, and more.

```
from flask import Flask

app = Flask(__name__)

@app.route('/')
def hello_world():
    return 'Hello, World!'

if __name__ == '__main__':
    app.run()
```

In the above code, we use Flask to create a simple web application that returns a "Hello, World!" message. We define a route using the @app.route() decorator, which maps the URL / to the hello_world() function.

Django

Django is a full-featured web framework for Python that is used by some of the largest websites in the world, including Instagram and Pinterest. Django provides a robust set of features for building complex web applications, including an ORM for working with databases, a templating engine for rendering HTML, and built-in security features.

```python
from django.http import HttpResponse
from django.shortcuts import render

    def hello_world(request):
        return HttpResponse("Hello, World!")

    def home(request):
        return render(request, 'home.html')
```

In the above code, we use Django to create two views: hello_world() and home(). The hello_world() view returns a simple message, while the home() view renders a template called home.html.

Pyramid

Pyramid is a flexible and scalable web framework for Python. Pyramid is designed to be modular, allowing developers to choose the components they need for their application. Pyramid supports a variety of databases and templating engines, making it a great choice for building complex web applications.

```python
from pyramid.config import Configurator
from pyramid.response import Response

def hello_world(request):
    return Response('Hello, World!')

if __name__ == '__main__':
    config = Configurator()
    config.add_route('hello', '/')
    config.add_view(hello_world, route_name='hello')
    app = config.make_wsgi_app()
    serve(app, host='0.0.0.0', port=8080)
```

In the above code, we use Pyramid to create a simple web application that returns a "Hello, World!" message. We define a route using the config.add_route() method and map it to the hello_world() function using the config.add_view() method. We then create a WSGI application using the config.make_wsgi_app() method and serve it using the serve() method.

Python provides a variety of libraries and frameworks for web development, each with its own strengths and weaknesses. Flask is a lightweight and flexible framework, Django is a full-featured framework for building complex web applications, and Pyramid is a scalable and modular framework. With Python, developers can build web applications quickly and efficiently, making it a great choice for web development.

Introduction to web frameworks (Flask, Django, Pyramid, etc.)

Web development is an essential aspect of modern-day technology. With the increasing demand for web-based applications, frameworks have emerged as a popular tool for web developers to create web applications with ease and speed. In this article, we will discuss the introduction to web frameworks, with a focus on Flask, Django, Pyramid, and other popular web frameworks.

What is a Web Framework?

A web framework is a software framework that helps developers create web applications by providing pre-built tools, templates, and libraries. Web frameworks help developers avoid repetitive tasks by providing reusable code blocks, so they can focus on implementing specific application requirements. The use of web frameworks reduces the development time, increases the productivity of developers, and ensures the maintainability of the codebase.

Types of Web Frameworks

There are two types of web frameworks:

Full-stack Frameworks: These frameworks provide all the components required to build a complete web application, including front-end, back-end, and database components. Django is an example of a full-stack framework.

Microframeworks: These frameworks provide only the bare minimum required to build a web application, and they leave the rest of the development up to the developer. Flask is an example of a microframework.

Introduction to Flask

Flask is a microframework for Python that is lightweight, easy to learn, and easy to use. Flask has a modular design and provides only the bare minimum required to build a web application. Flask is highly extensible and allows developers to use various extensions to add functionality to their application.

Flask uses the Jinja2 template engine to render HTML templates. Flask also provides support for database integration using various libraries like SQLAlchemy, Peewee, and Flask-SQLAlchemy. Flask supports different types of HTTP requests like GET, POST, PUT, DELETE, and more. Flask is suitable for building small to medium-sized web applications.

Introduction to Django

Django is a full-stack web framework for Python that is widely used in the industry. Django provides a robust framework for building web applications and follows the Model-View-Controller (MVC) architectural pattern. Django has a comprehensive set of tools that help developers create complex web applications quickly.

Django provides support for database integration using its Object-Relational Mapping (ORM) system. Django also has a built-in Admin interface that provides an easy way to manage data in the database. Django supports different types of HTTP requests and provides built-in support for authentication and authorization.

Introduction to Pyramid

Pyramid is a web framework for Python that is flexible, scalable, and highly extensible. Pyramid is based on the Model-View-Controller (MVC) architectural pattern and provides support for database integration using various libraries like SQLAlchemy, MongoEngine, and more.

Pyramid is highly configurable and provides a variety of configuration options for different use cases. Pyramid provides a minimalistic approach to building web applications and leaves most of the implementation details up to the developer. Pyramid is suitable for building small to large-sized web applications.

Web frameworks are essential tools for modern web development. Flask, Django, and Pyramid are some of the most popular web frameworks for Python, each with its own unique set of features and capabilities. When choosing a web framework, it is important to consider the requirements of the project and choose a framework that is suitable for the project's specific needs.

Using Flask for web development

Flask is a popular microframework for Python that is widely used for web development. Flask is lightweight, easy to use, and highly extensible, making it an ideal choice for building small to medium-sized web applications. In this article, we will provide an introduction to using Flask for web development, with a focus on its key features and how to use them.

Getting Started with Flask

To get started with Flask, you first need to install it. You can install Flask using pip, the package installer for Python, by running the following command in your terminal:

pip install flask

Once you have installed Flask, you can create a new Flask application by creating a new Python file and importing the Flask module. Here's an example of a simple Flask application:

from flask import Flask

```
app = Flask(__name__)

@app.route('/')
def hello():
    return 'Hello, World!'
```

In this example, we import the Flask module and create a new Flask application instance. We then define a route using the @app.route decorator, which specifies the URL path for the route. In this case, the route is /, which is the root path for the application. We then define a function that returns a string message when the route is accessed.

Running the Flask Application

To run the Flask application, you can use the flask run command in your terminal:

```
export FLASK_APP=hello.py
flask run
```

This will start a local web server on your machine that you can use to access your Flask application. You can access the application by navigating to http://localhost:5000 in your web browser.

Working with Templates

One of the key features of Flask is its support for templates, which are pre-built HTML files that can be used to generate dynamic content for web pages. Flask uses the Jinja2 templating engine, which is a powerful and flexible engine that provides a range of features for generating dynamic content.

Here's an example of using a template in Flask:

```
from flask import Flask, render_template

app = Flask(__name__)

@app.route('/')
def hello():
    return render_template('index.html', name='John')

if __name__ == '__main__':
    app.run()
```

In this example, we import the render_template function from the Flask module and use it to render an HTML template file called index.html. We also pass a variable called name to the template, which is used to generate dynamic content. We then run the Flask application using the app.run() function.

Working with Forms

Flask also provides built-in support for handling web forms, which are used to collect user input from web pages. Flask uses the request object to access form data, which can be accessed using the request.form dictionary.

Here's an example of using a form in Flask:

```
from flask import Flask, render_template, request

app = Flask(__name__)

@app.route('/', methods=['GET', 'POST'])
def form():
    if request.method == 'POST':
        name = request.form['name']
        return render_template('result.html', name=name)
    return render_template('form.html')

if __name__ == '__main__':
```

```
app.run()
```

In this example, we define a route that accepts both GET and POST requests. When the user submits the form, the POST method is used to send the form data to the server. We then access the form data using the request.form dictionary and use it to generate dynamic content in a template called result.html. If the request method is GET, we display a form for the user to fill out, which is defined in a separate template called form.html.

Working with Databases

Flask also provides support for working with databases, which are used to store and manage large amounts of structured data. Flask supports a range of different databases, including SQLite, MySQL, and PostgreSQL, among others.

Here's an example of using a database in Flask:

```
from flask import Flask, render_template, request
from flask_sqlalchemy import SQLAlchemy

app = Flask(__name__)
app.config['SQLALCHEMY_DATABASE_URI'] = 'sqlite:///test.db'
db = SQLAlchemy(app)

class User(db.Model):
    id = db.Column(db.Integer, primary_key=True)
    name = db.Column(db.String(80), nullable=False)

@app.route('/', methods=['GET', 'POST'])
def form():
    if request.method == 'POST':
        name = request.form['name']
        user = User(name=name)
        db.session.add(user)
        db.session.commit()
        return render_template('result.html', name=name)
    return render_template('form.html')

if __name__ == '__main__':
    db.create_all()
    app.run()
```

In this example, we first import the SQLAlchemy module and use it to create a new database instance. We then define a User model that defines the structure of the data to be stored in the database. We also define a route that uses the User model to add new data to the database when the user submits the form.

Using Django for web development

Django is a popular web framework for Python that is designed to help developers build complex, data-driven web applications quickly and efficiently. Django provides a range of features for handling common web development tasks, including URL routing, form handling, database management, and more.

Getting Started with Django

To get started with Django, you first need to install it using pip, the Python package manager. Once you have Django installed, you can create a new Django project using the django-admin startproject command.

```
$ pip install django
$ django-admin startproject myproject
```

This will create a new Django project with the name "myproject". The project directory will contain a number of files and directories that Django uses to manage your application.

Working with Models and Databases

One of the most powerful features of Django is its support for working with databases. Django provides a powerful Object-Relational Mapping (ORM) system that allows you to define your database schema using Python classes.

Here's an example of using the Django ORM to define a simple model:

```python
from django.db import models

class Book(models.Model):
    title = models.CharField(max_length=200)
    author = models.CharField(max_length=200)
    pub_date = models.DateField()

    def __str__(self):
        return self.title
```

In this example, we define a Book model that has three fields: title, author, and pub_date. The __str__ method defines how the object should be represented as a string.

To use the Book model in our application, we need to create a database table for it. We can do this using the migrate command, which creates the necessary database tables based on the models we have defined.

```
$ python manage.py makemigrations
$ python manage.py migrate
```

Working with Views and Templates

In Django, views are Python functions that handle requests and generate responses. Views can use templates to generate HTML pages dynamically.

Here's an example of a simple view that uses a template:

```
from django.shortcuts import render
from .models import Book

def book_list(request):
    books = Book.objects.all()
    return render(request, 'book_list.html', {'books': books})
```

Introduction to Django

Django is a popular web framework for Python that is designed to help developers build complex, data-driven web applications quickly and efficiently. Django provides a range of features for handling common web development tasks, including URL routing, form handling, database management, and more.

Getting Started with Django

To get started with Django, you first need to install it using pip, the Python package manager. Once you have Django installed, you can create a new Django project using the django-admin startproject command.

```
python
Copy code
$ pip install django
$ django-admin startproject myproject
```
This will create a new Django project with the name "myproject". The project directory will contain a number of files and directories that Django uses to manage your application.

Working with Models and Databases

One of the most powerful features of Django is its support for working with databases. Django provides a powerful Object-Relational Mapping (ORM) system that allows you to define your database schema using Python classes.

Here's an example of using the Django ORM to define a simple model:

```python
from django.db import models

class Book(models.Model):
    title = models.CharField(max_length=200)
    author = models.CharField(max_length=200)
    pub_date = models.DateField()

    def __str__(self):
        return self.title
```

In this example, we define a Book model that has three fields: title, author, and pub_date. The __str__ method defines how the object should be represented as a string.

To use the Book model in our application, we need to create a database table for it. We can do this using the migrate command, which creates the necessary database tables based on the models we have defined.

```
$ python manage.py makemigrations
$ python manage.py migrate
```

Working with Views and Templates

In Django, views are Python functions that handle requests and generate responses. Views can use templates to generate HTML pages dynamically.

Here's an example of a simple view that uses a template:

```python
python
Copy code
from django.shortcuts import render
from .models import Book

def book_list(request):
    books = Book.objects.all()
    return render(request, 'book_list.html', {'books': books})
```

In this example, we define a book_list view that retrieves all the books from the database using the Book.objects.all() method. We then pass the list of books to a template called book_list.html using the render function.

Here's an example of a simple template that displays the list of books:

```html
<!DOCTYPE html>
<html>
<head>
   <title>Book List</title>
</head>
<body>
   <h1>Book List</h1>
   <ul>
      {% for book in books %}
         <li>{{ book.title }} by {{ book.author }} (published on {{ book.pub_date }})</li>
      {% endfor %}
   </ul>
</body>
</html>
```

In this template, we use the Django template language to loop over the list of books and display the title, author, and publication date for each book.

Django is a powerful and flexible web framework for Python that provides a range of features for building complex, data-driven web applications. With its support for models, views, templates, and databases, Django makes it easy to create dynamic, interactive web pages that can be used for a variety of purposes. Whether you are building a simple blog or a complex e-commerce site, Django is a great choice for your next web development project.

Using Pyramid for web development

Pyramid is a web framework for Python that is designed to be flexible and easy to use. It provides a range of features for building web applications, including URL routing, templating, database integration, and more.

Getting Started with Pyramid

To get started with Pyramid, you first need to install it using pip, the Python package manager. Once you have Pyramid installed, you can create a new Pyramid project using the pcreate command.

```
$ pip install pyramid
$ pcreate -s starter myproject
```

This will create a new Pyramid project with the name "myproject". The project directory will contain a number of files and directories that Pyramid uses to manage your application.

Working with Views and Templates

In Pyramid, views are Python functions that handle requests and generate responses. Views can use templates to generate HTML pages dynamically.

Here's an example of a simple view that uses a template:

```
from pyramid.view import view_config
from myproject.models import Book

@view_config(route_name='book_list', renderer='templates/book_list.jinja2')
def book_list(request):
    books = request.dbsession.query(Book).all()
    return {'books': books}
```

In this example, we define a book_list view that retrieves all the books from the database using the request.dbsession.query(Book).all() method. We then pass the list of books to a template called book_list.jinja2 using the @view_config decorator.

Here's an example of a simple template that displays the list of books:

```
<!DOCTYPE html>
<html>
<head>
```

```
    <title>Book List</title>
</head>
<body>
    <h1>Book List</h1>
    <ul>
        {% for book in books %}
            <li>{{ book.title }} by {{ book.author }} (published on {{ book.pub_date }})</li>
        {% endfor %}
    </ul>
</body>
</html>
```

In this template, we use the Jinja2 template language to loop over the list of books and display the title, author, and publication date for each book.

Working with Models and Databases

Pyramid uses SQLAlchemy, a popular Object-Relational Mapping (ORM) library, for working with databases. To use SQLAlchemy in your Pyramid application, you need to configure a database engine and create a session factory.

Here's an example of how to configure a database engine in Pyramid:

```
from sqlalchemy import create_engine
from sqlalchemy.orm import sessionmaker

def main(global_config, **settings):
    engine = create_engine(settings['sqlalchemy.url'])
    session_factory = sessionmaker(bind=engine)
    return {'dbsession': scoped_session(session_factory)}
```

In this example, we create a database engine using the settings defined in global_config. We then create a session factory and return a dictionary that includes a dbsession key that is bound to a scoped session.

Pyramid is a flexible and easy-to-use web framework for Python that provides a range of features for building web applications. With its support for views, templates, and databases, Pyramid makes it easy to create dynamic, interactive web pages that can be used for a variety of purposes. Whether you are building a simple blog or a complex e-commerce site, Pyramid is a great choice for your next web development project.

Congratulations on getting this far, it shows that you really want to take your knowledge of Python to an expert level, our journey doesn't end here, this is volume number 1, stay tuned to find out when volumes number 2 and 3 come out, for complete the expert level.

I wish you success and hope to see you in the next edition, which will further deepen knowledge about Python, the full version will feature the union of Python and artificial intelligence.

Edson

About the author

Edson L P Camacho is a highly skilled professional with a degree in Technology in Digital Games and a postgraduate degree in Artificial Intelligence. With extensive experience in teaching and mentoring, he has helped hundreds of students to develop digital games using Unity and C#, as well as the Unreal engine.

His passion for learning and innovation extends beyond game development, as he is also a dedicated student of digital painting and 3D modeling for games. He continuously seeks to broaden his knowledge and expertise, ensuring that he can share only the highest quality content with his students.

Edson is a true industry expert, constantly pushing the boundaries of what is possible with cutting-edge technologies and techniques. His commitment to his students and to the field of digital game development is unparalleled, making him an invaluable resource for anyone looking to take their skills to the next level.

One day the prophet Isaiah said...

"All men are like grass and all their glory is like the flowers of the field... The grass withers and the flowers fall, but the Word of our God stands forever."

Isaiah 40: 7-8